What Scientists, Science Writers, and the Press Say about Dr. Shettles:

"In this...revision, some steps of sex selection have been refined or eliminated over the years, but the accurate determination of time of ovulation remains the basic tenet. The authors endorse the cervical mucus method for ovulation detection and consider other procedures as alternatives or additional means in achieving success. They attribute "traditionalists'" reticence to providing sex-selection guidelines to a concern for ethical questions of "gendercide," but they argue that most parents do consider sex selection to balance families. It is an alluring concept, judging by the popularity of previous editions. This updated and moderately priced version is highly recommended for public libraries."

—Mary Hemmings, Health Science Library,
McGill University, *Library Journal*

"A genius...ahead of his time."

—Albert Rosenfeld, former science editor of *Life* magazine

"His studies...are classic examples of exceedingly valuable research. He has contributed greatly to our knowledge of oocytes and spermatozoa...His *in vitro* fertilization of oocytes and culture of resulting embryos is a landmark in our insight into human embryogenesis."

—John Rock, M.D.,
one of the developers of the first birth-control pill

D0176525

"He developed a technique for transplantation of a fertilized egg into a fallopian tube. Called gamete intrafallopian transfer, or GIFT, the technique is considered one of the significant developments in infertility research. It is used today for couples who have been unable to conceive using conventional *in vitro* fertilization techniques."

—*New York Times*, February 16, 2003

"An ingenious mind ... superb technical ability."

—Dr. Alan F. Guttmacher, former national president
of Planned Parenthood-World Population

"One of the twentieth century's titans in the field of female infertility."

—Robert Weil, *Omni* magazine

How to Choose
the Sex of Your Baby

How to Choose
the Sex of Your Baby

The Method Best Supported by Scientific Evidence

Landrum B. Shettles, M.D., Ph.D., and
David M. Rorvik

Broadway Books
New York

PUBLISHED BY BROADWAY BOOKS

Previous editions of this book were published by Dodd, Mead & Co., New York, under the titles *Your Baby's Sex*, in 1970, and *Choose Your Baby's Sex*, in 1977. Further revised editions were published by Doubleday under the title *How to Choose the Sex of Your Baby* in 1984 and 1989, and by Broadway Books in 1997.

Published in the United States by Broadway Books, an imprint of The Doubleday Broadway Publishing Group, a division of Random House, Inc., New York.
www.broadwaybooks.com

Book design by Diane Hobbing of Snap-Haus Graphics

LIBRARY OF CONGRESS CATALOGING-IN-PUBLICATION DATA
Shettles, Landrum B. (Landrum Brewer), [date]
    How to choose the sex of your baby : the method best supported by scientific evidence / Landrum B. Shettles and David M. Rorvik.—[Rev. ed.]
        p.   cm.
    1. Sex preselection. I. Rorvik, David M. II. Title.
QP279.S5 2006
612.6—dc22

                                                                    2006043944

ISBN-13: 978-0-7679-2610-2
ISBN-10: 0-7679-2610-2

PRINTED IN THE UNITED STATES OF AMERICA

10   9   8   7   6   5   4   3   2   1

We dedicate this book to our readers and to couples everywhere who seek to choose the sex of their offspring. We dedicate this book, as well, to the growing number of doctors and researchers who share our belief that sex-selection technologies are, on balance, a force for good—helping to produce happier families and healthier societies.

Contents

A Note on the Collaboration

Lest the reader be confused by third-person references to Dr. Shettles, it should be understood that Mr. Rorvik is doing the writing and Dr. Shettles is providing the medical insight and expertise.

How to Choose
the Sex of Your Baby

Part One

Is It Possible?
Is It Moral?

Nearly 40 Years of Success!

Forty years! We find it hard to believe ourselves. For nearly four decades now, prospective parents from Biloxi to Bombay, Chicago to Cape Town, Seattle to Shanghai have been using our method to choose the sex of their children. *Millions* of people throughout the world have used the Shettles method with consistent good results, making us the past, the present, and, we trust, the future number-one guide to sex selection on the planet.

During those four decades, we have faced plenty of competition and quite a few "imitators" of our method. But as the competition has come and gone, the Shettles method has persisted, fueled primarily by satisfied word of mouth. Were it not for couples who have used our method with success and reported this to their friends and neighbors, we, too, would long since have fallen by the wayside. Failure doesn't keep anything going for years, let alone *40 years*. Only success can do that. And so, to you, our faithful readers, some of whom have used our method to attain not merely one child of the desired sex but two or more, often creating gender-balanced families in the process, we express our heartfelt thanks and appreciation.

And to those of you completely new to our book and our method, we say *Welcome*. We are delighted to add you to our ever-growing "family" and trust that you, too, will add to *your* families the children you will cherish, whatever their genders.

Since we last revised this book, some other sex-selection methods have come along, both low tech and high tech. We will be talking more about these later on. But, as usual, we have noted that when something "new" is announced in this field, it is often really just a restatement or a variation on components of the Shettles method.

By the same token, findings that sometimes claim to refute the Shettles method generally fail when they are more closely scrutinized or more thoroughly followed up over longer periods of time. We will provide examples of this later on. We will also tell you about some high-tech methods of sex selection, both old and newly emerging ones, that, unlike the Shettles method, are generally opposed by bioethicists and by the majority of doctors and medical professionals, for reasons we will discuss. We are confident that most couples will continue to find our approach to sex selection the easiest, the most natural, and the most reliable, as well as the most ethical.

This is the *sixth* revision of our book since it was first published in April 1970. It contains all of the latest sex-selection data. Our method has been consistently effective and consistently refined over the years to make it easier and more comfortable for all to use. *Our success rate continues to be 75 percent or better for those seeking girls and 80 percent for those who seek boys. And the rate of success is even higher among those who have reported to us on our questionnaires (see later in this book) that they were "highly confident" that they had precisely pinpointed the time of ovulation—a key factor in the Shettles method.*

In the pages ahead, you'll hear from a number of those who have tried the method—and we'll be answering questions many of you have sent us since our last edition appeared several years ago.

Again, congratulations for joining the sex-selection team that has been getting results in more than twenty countries for forty years. At a national meeting of the American College of Obstetricians and Gynecologists, Dr. Shettles was credited with having published "the landmark paper," in the early 1960s, which made sex preselection a subject that could and should be taken seriously. By having made the decision to investigate and, we hope, use sex selection yourself, you have become part of a landmark effort that we believe will continue to flourish.

Dr. Shettles' Track Record

Though the Shettles method remains "theory" and is disputed by some, you should be aware that Dr. Shettles has a formidable record for being both ahead of his time and *right*. He and Dr. John Rock of Harvard were the first to fertilize human eggs *in vitro,* launching what is today a revolution in fertility research. But it took literally decades for other researchers to follow up on their pioneering work. In the 1960s, Dr. Shettles discovered a method of obtaining fetal cells that could be used to assess fetal health and rescue distressed pregnancies, detect defects, and so on. Other researchers said they could not duplicate his research or simply ignored it—despite its enormous implications. Finally, researchers in mainland China reported they had duplicated the work, and this was then followed up on by researchers in Indiana, who finally realized the full import of the development and credited Dr. Shettles with its discovery. Today this technique (called chorionic villi sampling) has partially supplanted the more dangerous amniocentesis as a method of monitoring fetal health—and, unlike amniocentesis, it can be used from the very earliest stages of pregnancy with minimal invasion.

Then, in 1979, Dr. Shettles reported on another technique he had developed by which a fertilized egg could be surgically transferred directly into a woman's fallopian tube to achieve pregnancies that could otherwise not occur, owing to various infertility problems. At first, this technique, which has come to be known as gamete intrafallopian transfer, or GIFT—and it truly is a gift to many of the infertile—also was ignored and no credit was given. But GIFT rapidly became one of the crown jewels in the armamentarium of infertility research and treatment and is today regarded as one of the most important developments in that field in the twentieth century.

Finally, in 1991, in an editorial in the *Journal of in Vitro Fertilization and Embryo Transfer*, Dr. Shettles was given long-overdue credit and hailed as the father of GIFT. The editorial concluded:

> *In the case of GIFT a scan of the medical literature of the past 15 years clearly shows that L. B. Shettles should be credited with the introduction of the concept of transferring gametes into the fallopian tubes as a means of achieving a pregnancy. Frederick P. Zuspan, editor of the* American Journal of Obstetrics and Gynecology, *in his letter to Shettles notifying him of the acceptance for publication of his landmark paper, stated the need for its publication "as soon as possible as it opens new avenues for therapy."*

Dr. Shettles, who was once described by *Omni* magazine as "one of the twentieth century's titans in the field of female infertility," is delighted that he's still challenging orthodoxy right into the twenty-first century.

Can We Really Choose the Sex
of Our Children? *Yes!*

One of the reasons Dr. Shettles created the sex-selection method you are about to become familiar with is because so many of his patients at Columbia-Presbyterian Hospital in New York City asked him if there was any way he could help them tip the balance in favor of conceiving a child of a specified gender. Often these patients already had a child of one sex or more than one of the same sex and now, understandably, wanted a child of the opposite sex. Many asked the question with some hesitation, ultimately explaining to Dr. Shettles that other doctors had rebuffed them when they so inquired. Many were emboldened to ask Dr. Shettles this then-delicate question because he had been much in the news for his pioneering work in the field of female infertility. They imagined he might be sympathetic to their longings—and they were right!

Of course, they couldn't know at the time that Dr. Shettles had been asking himself the same question, a question that he would eventually answer not only to the satisfaction of his patients and hundreds of thousands of others around the world but also to his *own* satisfaction as he went about creating a family of his own, consisting of three boys and three girls!

Soon Dr. Shettles was not only able to tell couples who had not been able to conceive at all that, with techniques he helped perfect, they could indeed become parents but that it was also possible to choose the sex of one's children with a high degree of success. The Shettles method, as it has evolved, is *not* 100 percent successful. It is wise to keep that in mind at all times. But the method does very significantly increase your chances—elevating them from 50 percent if you do nothing—to better than 75 percent. And, in fact, as you will see later on in this book, some researchers have reported success rates with the Shettles method reaching *90 percent!*

What's the Evidence?

Next question: What evidence is there in support of Dr. Shettles' sex-selection theory? The evidence is of two kinds. One kind is "anecdotal" and consists of the reports of thousands of people who say they have successfully used the method. Anecdotal evidence is *not* scientific evidence, but it is often very useful and suggestive nonetheless. This is especially true when a significant number of the anecdotes issue from couples who have had three, four, or more children of one sex and then, upon first trying the Shettles method, finally have a child of the opposite sex.

The anecdotal evidence in support of the Shettles method is overwhelming. But it is not, to repeat, scientific evidence. A woman wrote a book several years ago proposing a sex-selection method that conflicted in many ways with that of Dr. Shettles. In some ways it appeared to offer *opposite* recommendations. (We will discuss that book in more detail later on, when we examine other sex-selection methods.) The point we wish to make here is that in the introduction to her book, this woman suggested—falsely—that we rely entirely on anecdotal evidence to support the Shettles method.

"The only way to determine whether a sex-selection method works," she wrote, "is, first, to enlist the cooperation of a large number of couples who state a sex preference, and, second, to note the sex outcomes of the pregnancies among couples who have meticulously followed the recommendations. . . . This is the *only* type of evidence that is convincing." We agree (and only wish there had been more of this kind of evidence in that writer's book). We will demonstrate, later in this book, that the Shettles method is better supported by precisely this kind of scientific evidence than is any other method in use today. In summary, it is the one sex-selection method *best supported* by the available scientific data, data that have been produced by independent researchers in numerous countries throughout the world.

It must also be pointed out that the Shettles method has persisted longer than any other sex-selection method in use today. Rather than lose followers and support, it has continued to gain them. That fact says something positive about the method. No sex-selection method can persist in the absence of "satisfied customers." Not only laypeople but an increasing number of medical and scientific researchers have come to discern merit in Dr. Shettles' sex-selection theories.

Dr. Shettles began developing his techniques in the early 1960s and has continued to refine them ever since. This is the sixth and by far the most comprehensive book we have written on the subject. We are confident that this book will make it easier than ever before for couples to utilize the Shettles method and, at the same time, to learn about other methods, some of which have at least partial validity and some of which, in our view, do not.

How It Began

How did Dr. Shettles get involved in sex-selection research? First, he did not enter the field with the idea of making abstract scientific points; rather, he was directly motivated by the desire to help alleviate some of the disappointment many of his patients expressed over being unable to conceive a child of the sex they desired. Time and again couples came to him with the same story: they had already had two, three, or more children of the same sex and very much wanted one of the opposite sex. In some cases it was clear that these couples would "keep trying" until they achieved their goal, even though they might already have had more children than they really wanted or could afford.

It became evident to Dr. Shettles, even many years ago, that a sex-selection method, if it could be made to work, would not only alleviate suffering among couples and within families but could also have a favorable impact on society as a whole. If couples could achieve sexually balanced families with a minimum of "tries," then there was a good chance, Dr. Shettles reasoned, that population growth, increasingly a threat to society, could be slowed down to some extent. Most parents and prospective parents told Dr. Shettles that what they had always wanted were two children—one of each sex. If they could have one boy and one girl, they said, they would consider their families complete.

Failure to achieve that ideal one boy/one girl balance, or at least to have children of both sexes, often resulted in psychological distress sometimes as acute, Dr. Shettles observed, as that experienced by some of his infertility patients—couples who had so far been unable to have children of *either* sex. As an authority in the field of infertility and human reproduction, Dr. Shettles was approached by many couples as their "court of last resort." Many came to him with woeful tales of indifferent and insensitive doctors who, in the case of the infertile, sometimes callously argued that "there are already too many babies, anyway," and, in the case of those with sexually imbalanced families, "you should just be happy you have children at all." Such arguments might have some validity in an abstract or general sense. But Dr. Shettles has never regarded his patients as either abstractions or generalities. They are individuals with individual problems and needs. Their problems require, and deserve, individual solutions.

Thus it seemed entirely appropriate to Dr. Shettles to try to apply some of the same ingenuity that he had used in solving various infertility problems to sex selection. He would try whatever seemed to work, guided by pragmatism. He did not worry about whether what he was doing fit into "accepted" or "appropriate" practice, as judged by his peers. If it was safe and effective, if it worked for his patients, then he was all for it. There is no denying that Dr. Shettles has raised a few scientific hackles in the course of his sex-selection work. There are some doctors, apparently, who think this whole field is beneath them. They refuse to acknowledge that the method works and will not investigate it themselves. Fortunately, there are more researchers who *do* have open minds, and, in any event, Dr. Shettles' reward has been the satisfaction of his patients and the many others who have made use of his work.

What's Ahead?

In the chapters that follow we will review a number of important issues and examine a variety of old and new findings related to sex selection. First, there is the crucial question: Even if we *can* select the sex of our children, *should* we? Is it moral? How do the various religions feel about it? What do the bioethicists say? What are the psychological implications—for the child as well as the parents? Is it really going to benefit society or will it pose new perils, producing, as a few have warned, an imbalance of males, with an attendant increase in violence, war, and so on? Are there certain types of people who should avoid sex selection? Are there medical reasons for using it?

After examining some of these issues, we will hear from some of the people who *have* used sex selection. Who are these people? What results have they reported?

In Part Two we explore early methods of sex selection, methods that have undergone gradual modification as knowledge of human reproduction has increased. We review the processes of conception as we currently understand them; this will provide the basis for comprehending the early work that led to Dr. Shettles' sex-selection methodology. We also track the evolution of Dr. Shettles' techniques, with emphasis on refinements and alterations not present in our previous books. We emphasize the growing body of scientific evidence that helps confirm the validity of Dr. Shettles' findings. Also, we discuss some other sex-selection methods that have emerged in recent years and explain why we believe these other techniques either have or do not have validity.

Part Three presents the specific instructions for selecting the two sexes. Charts are included to help guide you in the successful use of the technique. The most commonly asked questions about the procedures are answered. Environmental and hereditary variables that may influence the outcome of sex selection are discussed.

Finally, in an Afterword, we look ahead to probable near and more distant developments in sex selection. We examine some experimental methods already in early use.

A Questionnaire is provided at the end of the book for you to fill out after you have completed your attempt. We hope you will take time to complete the questionnaire. In doing so you will be helping us to better help others, in future editions of this book.

For Your Best Chances of Success...

IMPORTANT: Be sure to read the *entire* book before making your sex-selection attempt. The commonest cause of failure is impatience. Usually, when we get a questionnaire that reports an unsuccessful attempt, we are able to find evidence of either obvious carelessness in some particular or of failure to follow instructions properly. Most readers *do* follow the instructions carefully, thereby increasing their chances of success. We hope you will be among them.

To be conservative, we have stated that the overall success rate for the Shettles method is *at least 75* percent. This assertion is based on Dr. Shettles' own experience as well as that of other researchers who have conscientiously used the method in their practices. Couples who adhere closely to his instructions and who are accurate in their timing, Dr. Shettles is convinced, will, when using the "girl method," succeed 75 to 80 percent of the time; when using the "boy method," they will, under the same circumstances, succeed 80 to 90 percent of the time.

Is It Moral?
Should We Do It?
Will the World Be Overrun
by—Gasp!—Males?

Frankly, after nearly 40 years of witnessing the overwhelmingly benign results of sex selection, we don't know whether to laugh or cry when we see some of the apocalyptic visions conjured up by those who foresee dire consequences of this technology. One of the most absurd surfaced recently in a supposedly scholarly book called *Bare Branches: The Security Implications of Asia's Surplus Male Population.* Beginning in 2004, the Internet presented one of the more sensational speculations of this book, which has been repeated elsewhere since. The authors of this book argue that the rising popularity of sex selection, particularly in Asia, will result in a horrendous imbalance of surplus males. Ronald Bailey, in a column at reasonline.com, noted that the authors of *Bare Branches* envision "a hoodlum army of 30 million single men that by 2020 will be a menace to world peace." Uh-oh! Time to man—or, more likely, *woman*—the barricades and begin guarding against all this male excess brought on by runaway sex-selection methods.

Or is it?

Our first reaction to this new dark prediction was: Where have we heard this before—and how many times? *Many* times. And from some surprising sources. For example, Dr. Amitai Etzioni, a well-respected sociologist at Columbia University, once suggested that sex selection could result in "an over-production of boys" and that this excess male surplus would "very likely affect most aspects of social life." If sex selection became widespread, he warned, the results could be everything from the downfall of the two-party system to a very significant increase in the incidence of male homosexuality, prostitution, and violence—all due to there being too many males and too few females.

What is so galling about these sensationalist predictions is that they have no real basis in fact. On the contrary, they are directly contradicted by the facts. And even if it were true that people have a strong preference for male children, would the predictions of some authors that an excess of males will result in more humiliation and rapes of women come true? No, argue such authors as Gita Sen and Rachel C. Snow in *Power and Decision: The Social Control of Reproduction*. They reasonably suggest that a decrease in the number of females would instead result in greater value being placed on them and thus greater respect being paid them by competing males.

Do People Really Prefer Boys to Girls?

Nature has seen fit to produce a gender balance of about 105 males for every 100 females. There may be some good reasons for this slight imbalance since men, the weaker sex in terms of longevity, die off sooner. So there is a slight "natural" bias in favor of more males—at least at the outset of life—but what about the "unnatural" biases at work in the world? It is true that in some poor and emerging countries, where male labor is still highly useful for basic survival, a premium is placed on male offspring. In some places, abortion of unwanted genders and even infanticide have carried over into the present day. But social and economic trends ultimately defeat these biases. The governments of both China and India, for example, have now made it illegal to use abortion as a sex-selecting methodology and campaign actively to create more gender-balanced populations. As those economies improve—and they have improved enormously in recent years—government efforts are paying off.

In the West, recent surveys show that some populations favor fe-males over males, quite the opposite of what we have been told to ex-pect. But, overall, the trends are running strongly in favor of *balanced* families consisting of an equal number of boys and girls, usually one of each. Recently German bioethicist Edgar Dahl of the University of Giessen, addressing the Royal Society in London at an international conference on assisted human reproduction, cited studies showing that in the United States, Germany, and the United Kingdom, all countries where sex-selection technologies and methodologies are more readily available than in most other parts of the world, there is *no notable gen-der preference overall.*

It *is* true that there is some preference that the firstborn be a male. This preference is actually higher in the United States than in most European countries. Professor Dahl cited data showing that about twice as many Americans prefer that their firstborn be a boy rather than a girl. But then, like their European counterparts, they want their sec-ond child to be of the opposite sex. Thus, in the West, Professor Dahl concludes, "There is no evidence at all that there is a threat to the sex ratio."

In our own most recent survey of those using the Shettles method, we found that those who notified us in advance of their intentions to try for either a boy or a girl were *evenly divided* in preference between boys and girls. Moreover, in almost every case, those wanting a boy al-ready had one or more daughters, and vice versa. In those cases where couples stated they were using sex selection during their *first* efforts to conceive, we did find a slight preference for male offspring as first-borns.

It was very rare to encounter a couple who already had one child and who wanted, as a second child, offspring of the same sex. It was rarer still to find those with two or more children of the same sex wanting a third child of that gender. When we did come across the latter situa-tion, there were usually peculiar reasons for it. One woman, for exam-ple, told us that "I love babies and want more, but my nerves just won't tolerate boys." Another couple told us they wanted a fifth boy "because we're just a very macho kind of family." Almost everybody else wanted a family evenly divided between male and female offspring. Most wanted one of each.

Some commentators who are opposed to sex selection have pointed with alarm to reports that came out of China in late 1982. Headlines like this one cropped up everywhere: "Rural Chinese Reported Killing Their Baby Girls." There *is* still a strong pro-male bias in countries like China where rural people, in particular, view sons as economic assets and a form of security; sons can work in the fields, carry on the father's work, and support the parents in their old age. But what often was overlooked in these reports of female infanticide is the fact that China is attempting to impose a strict policy of population control that permits only *one* child per family. The killing of female children therefore is not so much an expression of anti-female bias as it is an expression of the desire to have at least one son. Any instance of infanticide is appalling, of course; if sex selection was widely available in China, we can only conclude that the frequency of infanticide would be reduced. And if the policy of the country permitted two children rather than one, it seems likely that many Chinese would still have—and *want*—daughters.

The most definitive study to date on what would happen if sex selection came into wide use in the United States was carried out by Dr. Charles F. Westoff of Princeton University's Office of Population Research and by Dr. Ronald R. Rindfuss of the University of Wisconsin's Center for Demography and Ecology. Detailed surveys of some six thousand married women revealed that an excess of male births would, in fact, result, but this excess would prevail for only two years, during which time more childless couples would choose to have sons as their firstborn. Thereafter, these same couples would have daughters, and a balance, the study indicated, would be restored. Other researchers have confirmed these findings.

This desire that the firstborn be a son is, nonetheless, worthy of note. Why this desire persists, even in developed countries, no doubt has something to do with the diehard desire of the husband to perpetuate "the family name." There also seems to be a perception that sons may, even in technologically advanced societies, somehow come in more "handy" and/or somehow be more economical, though a recent book indicates there really isn't that much difference in the cost of raising sons and raising daughters ($226,000 for the son versus $247,000 for the daughter, on average).

In 1987, Dr. Shettles was asked to debate the issue of sex selection in the "Point/Counterpoint" column of *Physician's Weekly*. Dr. LeRoy Walters, director of the Center for Bioethics at Georgetown University, argued that even if the number of male and female offspring balance out over time, the preponderance of sons among firstborn children could have adverse effects. "If there's an advantage to being a first child, it would accrue mainly to boys. . . . also, making a big issue of a child's sex reinforces stereotypes and perhaps contributes to sex discrimination." So Dr. Walters' conclusion was that sex selection should be discouraged, "especially for firstborn children."

Dr. Shettles countered that there is no evidence whatsoever to indicate that a preponderance of firstborn male children will have any adverse effects on society. Nor can he find any compelling reason to believe that sex selection will in any way contribute to sexual discrimination. Indeed, it may have the opposite effect. Couples' desires for firstborn male offspring exist independent of sex-selection technology. If couples do in fact desire sons as their firstborn children and do not get them, it could be argued, they may, however unintentionally or subtly, exhibit their dissatisfaction in ways that will eventually become apparent to (and harmful to) their daughters.

In addition, Dr. Shettles' own extensive experience with sex selection has convinced him that couples value daughters just as much as sons. "Never, during all these years, have I seen the sex ratio take off on a tangent that would yield multitudes of males. Even in China, where the government is trying to limit population growth with a one-child-per-family policy, they know that in order to bear children they're going to need females."

At the subsequent national meeting of the American College of Obstetricians and Gynecologists, the debate over this issue continued. Dr. Walters again voiced his concerns. He acknowledged that there is no legal way to prevent couples from using sex-selection technology, noting that "the constitutional right to privacy in connection with re-production is so strong that I think it would withstand any state-sponsored ban on sex preselection." Presumably, Dr. Walters supports that constitutional right. But, on a social rather than legal level, he nonetheless urges prospective parents to "give female children an equal chance to suffer from the disadvantages or have the privileges of first-bornness."

Bioethicist John C. Fletcher, speaking for himself, rather than for the National Institutes of Health, where he works, is one who has reversed his position on sex selection and now opposes it as "inherently sexist" because "there is a universal preference for males." The late anthropologist Margaret Mead differed, however. She strongly favored the use of sex selection because, she argued, "for the first time in human history, girls would be as wanted as boys." What she meant is that if a girl was the product of sex selection, she could grow up secure in the knowledge that she had truly been wanted and was not simply the child her parents had "settled for" when they failed to have a son.

More recently, a newer study has again confirmed that many of the fears about sex selection are ill-founded. Specifically, demographers Paul Schollaert of Old Dominion University in Norfolk, Virginia, and Jay Teachman of the University of Maryland, after analyzing the child-birth records of more than two thousand couples who had children of only one gender, found both groups (boy-only/girl-only families) equally intent on having a child of the opposite gender. There was no bias in favor of boys. The desire today, one of the researchers said, is for "a matched pair."

Also encouraging is a *New York Times* News Service report that "in the United States, doctors interviewed around the country say that abortions on the basis of sex are extremely rare."

Psychological Advantages of Sex Selection

This brings up an important point—and one that argues powerfully for sex selection, whether it is used to conceive sons *or* daughters. Many children of both sexes have had to suffer severe and enduring psychological trauma as a result of perceiving—often correctly—that they were not what one or both of their parents wanted. No matter how hard a parent will try to hide his or her disappointment over a son whom that parent had hoped would be a daughter, or vice versa, that disappointment will almost always be communicated in various ways to the child, who will then grow up feeling he or she has somehow "failed." Withdrawal, antisocial behavior, or gender confusion can result.

Dr. Thomas Verny, in his groundbreaking book, *The Secret Life of the Unborn Child,* reported on a pilot study he conducted in which he sought to determine the effects prenatal events and birth experiences seemed to have, in later life, on a group of individuals who were in psychotherapy. The Toronto psychiatrist concluded that how parents felt about their children at birth and even before birth could have lasting impact and be predictive, in some important respects, on the future behavior and mental health of those children. Two essential elements for the development of a healthy personality, he concluded, are a positive attitude toward the pregnancy on the part of the parents and the fulfillment of their wish for a child of a desired sex.

In other words, the two most important perceptions we can have of ourselves is that, first, we are wanted as human beings and, second, that we are wanted, specifically, as sons *or* daughters. "In both men and women," Dr. Verny states, "that combination produced less depression, less irrational anger and better sexual adjustment."

Another medical researcher, Dr. A. L. Benedict, has suggested that some individuals are really only suited for raising children of one sex. Sex selection could be used in such instances, he added, to help ensure that neither those parents nor any offspring suffer undue trauma. This idea makes us a bit uneasy, however, because a parent who cannot adapt at all to children of one sex must almost certainly be psychologically disturbed to the point where it would be better for him or her to have no children. But since that individual is likely to go ahead and have children, no matter what *we* think, it can be argued that sex selection could be useful even in such extreme cases.

Are *You* Suited for Sex Selection?

Dr. Shettles is always on the lookout for couples who would be badly disappointed if an attempt at sex selection fails. And since current methods are *not* without failures, we constantly caution that no effort should be made to have a child, using any sex-selection method, unless the parents are quite sure they will be happy with a baby of either sex. Here is an example (a composite) of the sort of letter we occasionally receive that puts us immediately on guard:

Dear Dr. Shettles:

My husband and I have three beautiful daughters, and though we adore each of them and would not give them up for the world, I must confess that we—and my husband in partic-ular—have wanted a son for so long. After the last daughter was born, my husband couldn't hide how he felt and, for that matter, neither could I. Though we love this last baby as much as the others, we just couldn't help feeling terribly depressed. It all seems so unfair.

We couldn't even think about, much less talk about, having another baby—not until we heard about your sex-selection work. We've been smiling ever since and are now planning to "try again." We're just about ready, but we were wondering if there are any last-minute developments we should know about, as we want to be absolutely certain that we'll get a boy this time. . . .

Sincerely,

Mr. and Mrs. Anxious

The emotional extremes revealed in this letter are not common. Most people who write for further advice are realistic and are quite ca-pable of establishing a loving relationship with a child, no matter what its sex. A letter like the one above, however, will get a response some-thing like this:

Dear Mr. and Mrs. Anxious:

We must congratulate you, first of all, for having conceived three healthy daughters. At the same time, we sympathize with you for the disappointment you have felt in not having a son. We wish that we could recommend that you "try again," using our techniques, but we cannot.

As you know, if you have read our book carefully, these methods are *not* fail-safe; there is a chance you will not succeed. And statements in your letter force us to the conclusion that a "failure" might result in extremely serious depression. Unless or until you are honestly convinced beyond any doubt that you can have another child—of either sex—without experiencing anguish or depression or other emotions that will be harmful to yourselves, your marriage, your new baby, or your other children, we urge you *not* to conceive another child. Perhaps you can resolve some of the conflicts you feel through professional counseling. . . .

In short, sex selection is not for everyone. But for the many who are starting their families or who are determined to have another child and are sure they will love it, regardless of its sex, the methods described in this book provide a means of influencing factors toward the desired end. To demonstrate how different attitudes can be, consider this letter from a couple with three girls. In this case we felt no reluctance in recommending sex selection.

Dear Dr. Shettles:

We have just read about your work and are very interested in using it for our next baby. We have three daughters but had decided to have a fourth child even before we read your book. We have always wanted four children. Naturally, we would have preferred that two be boys and two be girls, but even if our next baby is another girl we will be happy and consider our family complete. Since your method seems easy to follow, we

figure we have nothing to lose in trying for a boy. As my monthly cycle is irregular, however, I'm writing to find out if you can tell me . . .
Sincerely,
Mrs. J.A.

As indicated, Dr. Shettles was not reluctant to advise Mrs. J.A. The attitudes revealed in this letter disclose a healthy outlook and a good prognosis for all concerned—including the new baby, whatever its sex.

What Churches and Other Authorities Say

Some readers inquire about the religious and legal status of sex selection. There is no legal prohibition on sex selection in the United States and in most parts of the world. Certainly there is no prohibition on the Shettles method. In some countries, high-tech methods of sex selection are prohibited by law. In the United Kingdom, for example, couples who seek to overcome infertility through implantation of embryos are not allowed to "sex" those embryos, which is to say they cannot select embryos on the basis of the gender of those embryos. We have more to say about this later on in the book when we discuss some of these emerging high-tech methodologies.

As for bioethicists, religions, and professional medical organizations, opinions of sex selection vary enormously. We certainly do not regard sex selection as immoral or sacrilegious in any way. Many Protestant ministers have used our methods in their own family planning over the years and have expressed no qualms about doing so. Similarly, rabbis have cooperated with Dr. Shettles in his research, and the Catholic Church, through Monsignor Hugh Curran, when he served as director of the Family-Life Bureau of the Archdiocese of New York, declared that the church has no objections to the Shettles method or to sex selection in general "as long as the intent of these efforts is not to prevent conception" or to harm or destroy embryos. Unlike some of the high-tech methods of sex selection that are emerging, the Shettles method poses no peril whatever to embryos.

We certainly oppose some *postconception* methods of sex selection, and we wish to make that very clear. We *do* regard as immoral those methods, for example, that identify embryos as being of one sex or another and then destroy those of the "wrong" gender. Abortion for this purpose is absolutely unjustified in our opinions.

Meanwhile, various professional medical organizations are taking different views on sex-selection technologies. One of the major organizations, the American Society of Reproductive Medicine, has generally taken the view that preconception sex-selection methodologies that are safe and effective are acceptable, even for nonmedical reasons. The society, which is made up of doctors, nurses, and scientists, places value on the psychological well-being that it believes accrues from family balancing.

Sex selection may eventually acquire virtues that may not be obvious today. If fail-safe preconception sex-selection methods should emerge, they could be used to prevent the births of children who carry sex-linked genetic disorders. Only males, for example, suffer from hemophilia, the "bleeding disease." The recessive gene that is the cause of this disease can express itself only in males. Known carriers of the disease could stop passing the gene on to future generations by avoiding male conceptions. Many other sex-linked diseases could also be avoided in this fashion. "The elimination of these disorders in one generation, by a judicious choice of the sex of the offspring," Drs. Robert Edwards and Richard Gardner have written (in the journal *New Scientist*), "would not only be of direct benefit to that generation, but would benefit the race for generations to come."

Sex selection of any sort is not without some social risk. But that risk appears to us to be minimal and is easily outweighed by the potential benefits, which include satisfied parents, happier, healthier children, and smaller, better-balanced families.

Letters from Home
and Our Results to Date

Who would want to choose the gender of their offspring?

A very large number of people, it turns out. Over the nearly 40 years this book has been in print, we have heard from all manner of people and from nearly every geographic region in the world—from bankers and plumbers, nurses and farmers, engineers and teachers, medical doctors and landscapers, from large cities all over and remote villages in faraway countries, from dwellers of high-rises, jungles, and even desert islands. We've heard from people in war zones and even from one couple who reported using the method during a volcanic eruption!

Do Even Princes Use the Shettles Method?

Occasionally we hear from people most of us would regard as rather "exotic." Over the years, Dr. Shettles was consulted by several prominent people in the entertainment industry, including a couple of actresses who are household names, by Arab and Indian potentates, and by members of royal families. One gentleman, from a counry that is either benighted or enlightened, depending on your point of view, had children with three different wives, using the Shettles method. What made this remarkable was the fact that he was married to all three at one time!

Dr. Shettles did *not* advise Prince Charles and the late Princess Diana, although rumors and some news reports circulated that the royal couple studied the Shettles method with the goal of begetting a son. No secret was made of the fact that they—and especially Prince Charles—wanted a son. And it was apparently true, as widely reported in the press, that Prince Charles did a lot of reading on matters related to conception and pregnancy. One news account about this activity mentioned the Shettles method and an earlier edition of this book. The couple, of course, did in fact go on to conceive sons, but, without more information, we can't claim any of the credit. Still, we admit to being intrigued.

Most of the people we hear from are "ordinary folks"—often young couples just planning their families or hoping to complete them with one more child of the "other" gender, to achieve a gender-balanced family. We hear from professional women in many fields—and we also hear from others we also consider professionals: housewives. Sometimes husbands and prospective husbands are the ones to contact us first. Quite a number are in the medical professions, including psychiatry.

Quite often we hear from couples who have used the method with success against overwhelming odds. Consider this testimonial, for example, which was published in an issue of the magazine *Family Circle*:

> *A big thank you for having taken time to send me a photocopy of an article . . . "Now You Can Choose the Sex of Your Next Child" by David Rorvik and Landrum B. Shettles, M.D. I am happy to announce that it worked. . . . My husband and I are the proud parents of the first girl in the Harrison family in 250 years.*

Over the years, a number of people from families with long histories of producing nothing or almost nothing but boys have consulted Dr. Shettles. While they might be envied in some third-world countries, they are usually quite unhappy here. This tendency to have nothing but male offspring is no doubt rooted in genetics to some extent, but, even then, we have found that most of these men have at least some female-producing sperm. And though these may be far outnumbered by the male-producing sperm, the Shettles method can help make the conception environment more favorable for them, allowing, as in the case just described, for a "breakthrough" daughter.

Report from the Home Front: 1970 to 2006

(Some Unedited Commentary, Both Favorable and Unfavorable)

Most authors who write about sex selection, just like those who write diet books, usually include testimonial letters to support their ideas. We have always done this in past editions, and this section of the book has always been a favorite with our readers. We are aware, however, that many authors "cherry-pick" their testimonials with great care—that is, they cull out the unfavorable comments and present only those that support their theses.

Do we like running negative letters about our method? No. But, as you will see, the positive letters *far* outnumber the negatives. And, in any case, we have always acknowledged that the method is not fail-safe but does increase chances for success very substantially. Later on in this chapter we will tell you why we believe most people succeed but also why we believe some do not.

Let us begin, however, with some revealing letters from much earlier editions. We include them not only to help document the anecdotal success of our method (we talk about scientific documentation later) but also to help you understand why and how various couples used the method, sometimes to overcome difficult odds.

Letters from Home

Probably the best way of letting you know "who does it," and with what results, is to give you a sampling of the letters we have received, dating back to 1971 and coming up to the near present. We'll get to the *scientific* results in a subsequent chapter. (Names, initials, and addresses have been changed to protect confidentiality.)

Sept. 29, 1971
Dear Dr. Shettles:
   Thank you! After many years of questions and four female

children later, your theory has worked for our benefit. . . . The addition of our new son has indeed changed our lives and has convinced us that you have found a solution to the problem of repetitious sex of children among parents looking for a balance in females and males.

Most appreciative, W.F.M., Philadelphia, Pa.

May 21, 1972
Dear Dr. Shettles:

I read and studied your book and . . . conceived using your method of obtaining a female. Happy to say, on August 7 my healthy daughter was born—much to the surprise of everyone on my husband's side, since I have a two-year-old son, and no girls have ever been recorded on my husband's side for many generations. Thank you for all your help.

Sincerely, J.J., Buffalo, N.Y.

June 12, 1973
Dear Dr. Shettles:

I just wanted to take this opportunity to tell you of our experiences after reading your book. I am the proud father of three girls. For my fourth and final chance to have a little boy, my wife and I purchased a copy of your book. We followed it intensively and we had a beautiful healthy baby boy. We loaned your book to one of our friends, who is a prominent ophthalmologist. He happens to be the father of three boys. He also followed your book and *voilà!* he fathered a baby daughter. You are batting 1000. Keep up the good work! Many thanks.

Very truly yours, M.C., Joplin, Mo.

October 19, 1974
Dear Dr. Shettles:

I wish to thank you for sharing your scientific and medical knowledge with the public. My husband and I have a new lit-

tle son in our family of four daughters. We carefully carried out
your new directions on how to have a boy.
Sincerely yours, C.P., Sausalito, Calif.

October 16, 1975
Dear Dr. Shettles:

Please include us in your list of successes; we got a child of
the sex we desired. After having had a boy we wanted a girl,
and B. was born earlier this year. . . . We found the correct day
of ovulation and ceased having intercourse two days prior to it.
As prescribed, we used the douche, I had no orgasm, and shallow penetration and the missionary position were employed.
Thanks so much for what you have done to help people
like us.
Dr. and Mrs. J.C., New York City

January 22, 1976
Dear Mr. Rorvik:

Last Thursday I called Dr. Shettles to inform him of the
birth of my second son on January 14, and he suggested that I
write you a note and give you the information that my wife
and I used his method of abstinence, baking-soda douche, and
intercourse all on the day of ovulation in order to increase our
chances of having a son. After we had three daughters in a row,
my wife . . . talked with a researcher at the University of
Michigan who informed her that Dr. Shettles was doing some
work on selecting sex. . . . In 1970 we took a trip to New York
and met with Dr. Shettles and shortly thereafter we began using his method. Our first son was born in March of 1971.
Then last week our second son was born. In looking back and
reviewing in our minds the conception of our three daughters,
it was easy to understand how the chances of their being
girls . . . were greater.
Sincerely, Dr. M.V., Detroit, Mich.

September 16, 1978
Dear Dr. Shettles:

After following your procedures for conceiving a boy, and after three daughters, I gave birth on August 17 to William. My husband and I both want to thank you for your book and the precise directions. Thank you!
Sincerely, Mr. and Mrs. A.A., Palo Alto, Calif.

[From a completed questionnaire returned to us, dated February 7, 1979:] This is the first girl born to my husband's family in 42 years. My husband had no sisters or nieces. We thank you so much. We are ecstatic with our girl. We never thought we would succeed with my husband's family history. . . .
Mrs. L., Denver, Colo.

February 17, 1979
Dear Mr. Rorvik:

We are very, very happy to have been successful in our attempt for a boy . . . our family is now complete! In an age where family size is necessarily limited, I believe this method can be successfully applied to help families attain the wanted ratio of boys and girls. Most people, I believe, want a child of each sex. . . . I find nothing offensive about these methods and believe that they can definitely help to ensure that every child born is wanted and desired in every way. . . .
Mrs. J.B., Montreal, Quebec

May 27, 1979
Dear Dr. Shettles and Mr. Rorvik:

We are so happy we finally have our little girl. After having two boys words cannot describe what I felt when the doctor said, "It's a girl!" I couldn't believe it. All I could say was, "Really! I want to see! I want to see!" Everyone in the labor and delivery

room knew that I had two boys and had tried your method. I think they were almost as anxious as we were to find out what it was going to be. When we did, there was hugging and kissing all over the place. I even remember kissing the anesthesiologist.
Mrs. K.C., Pomona, Calif.

[From a completed questionnaire returned to us, dated August 27, 1979:] My relationship with my mother is so special to me that I feel a mother-daughter love is an unequaled privilege. I only hope I can guide my [newborn] daughter as fairly and lovingly as my mother guided me. As I sit here and look at my beautiful daughter, I thank you.
Mrs. L.S., Honolulu, Hawaii

[From a completed questionnaire returned to us, dated February 11, 1980. The writer, a dentist, reported a successful attempt to have a son, after having had two daughters.] The [baby] doctor appeared mostly indifferent and skeptical [about the sex-selection attempt]. When we bore the correct sex, I promptly reminded him of the procedure we used. He felt it was luck. We don't agree! Whatever you do, disregard the intimidating comments of your fellow doctors in the profession. One thing we've noticed is their lack of commitment and willingness to stick their necks out in believing in a possible answer to this problem. The least they could do is direct us to your book and leave it up to us to judge (and follow the suggestions). We've told as many people who are interested, as possible, in hope they will be as pleased with the results as we were. Thank you again for writing the book.
Dr. V.N., St. Louis, Mo.

May 24, 1981
Dear Dr. Shettles:

People laughed at my husband when he told them, practically from the day I conceived, that we had a girl. No one laughed

louder than his father, who has seven sons and one daughter. We followed your book exactly, and I got pregnant with our first attempt. Our daughter is twelve days old . . . we're thrilled with what we've accomplished. I wish all couples who are after a planned family could read your book. I also wish doctors would accept your research and be supportive . . . thank you again.
Mrs. H.D., Tucson, Ariz.

[From a completed questionnaire returned to us, dated August 18, 1982. The writer, an M.D. specializing in obstetrics and gynecology, reported success in conceiving a boy, after having had two girls.] Our home is filled with happiness because we have the baby boy that we've been dreaming of. . . . Thank you so much for having worked very hard to come up with this sex-selection procedure. . . .
Dr. and Mrs. R.O., Los Angeles, Calif.

March 12, 1983
Dear Dr. Shettles:
    Two boys were enough. I told my husband that since he wanted another child so badly, we would have to try for a girl. He was quite astonished when he found I'd already bought a book on the subject—yours! He is a research scientist and was very skeptical at first—but changed his mind after *he* read the book himself. Then he went from belittling to badgering— badgering me to do everything exactly "by the book." We did—and the result is our beautiful new daughter.
Mrs. R.S., Merion, Pa.

October 7, 1984
Dear Mr. Rorvik and Dr. Shettles:
    Your book not only helped us get the son we wanted but also helped us get pregnant. We had been trying for three years without success, despite help from several doctors. Your book was so clear and helpful; it enabled us to understand the reproduc-

tive system better than anything any of the other doctors told us. We would have been happy with any baby but we're doubly delighted that we got a son. We used your instructions for finding ovulation time and came in right on target. The combination of using the CM method and the BBT really helped.
Mrs. C.J., Fargo, N.D.

June 12, 1985
Dear Dr. Shettles:
    Thanks for an excellent, well-researched, easily understood book. We've recommended it *numerous* times! It really works!
Mr. and Mrs. K.L., Astoria, Ore.

[From a completed questionnaire returned to us, dated September 10, 1985:] We are so happy with our family and believe your method helped us achieve our daughter. We also believe the human body was designed by God and that he has allowed you to put these facts to use to come up with your methods. Thank you for doing this research.
Mr. and Mrs. V., Cicero, Ill.

[From a completed questionnaire returned to us, dated January 5, 1986. The writer stated that she and her husband had previously tried sex selection using the method recommended by Elizabeth Whelan—discussed later in this book. The Whelan method is, in many ways, the exact opposite of the Shettles method. As this couple's first child was a girl, they wanted a boy for their second child. Using the Whelan technique, however, they got a girl. For their second attempt to conceive a son they turned to the Shettles method and succeeded.] Since our success, I've told many friends about your book. One couple has already tried the method—for a girl—and succeeded. My sister used your book twice and has two sons—both planned. Thank you.
N.B., Toledo, Ohio

March 3, 1986
Dear Dr. Shettles:

We read your book very carefully and the result—a boy—was marvelous. We did not think we would ever have one after having had three daughters.

Mr. J.M., Mexico City

In 1986 we received a letter from a woman in Florida who stated that she had *five* sons, none via sex selection, and was now pregnant again, having this time used our method to try to conceive a daughter at last. She wrote that she was *very* confident she had done everything correctly and felt with great certainty that she would give birth to a daughter in 1987. She added that, of course, if the baby turned out to be a boy again she would love the child "100 percent" nonetheless. Still, whenever we get letters like this—and we get a fair number of them—we hold our breath along with the prospective parents and wait anxiously for the results. Fortunately, more often than not we are relieved by the outcome—and this time was no exception. In mid-1987 we received a completed questionnaire from this woman announcing "our" success—a daughter. She noted that her obstetrician had recommended the Shettles method.

February 14, 1987
Dear Dr. Shettles and Mr. Rorvik:

Shortly after the birth of our second child, another boy, my mother handed me a magazine article about your sex-selection method. I was intrigued and promptly went out and purchased *How to Choose the Sex of Your Baby.* I also read *Boy or Girl?* by Elizabeth Whelan, Sc.D., and *The Preconception Gender Diet* by Sally Langendoen, R.N., and William Proctor. I concluded that the Shettles method made the most sense, especially when compared with my own personal experience. Both my boys were conceived very close to ovulation. I knew this because I used a natural family planning method of birth control that utilized BBT and CM observations to determine my fertile period and thereby avoid it. By the same token, when I decided to get pregnant I knew my best chances would be close to ovulation, so I was able to time it within 24 hours. I did this by having intercourse on the day that I had "peak" CM symptoms. Unknowingly, I was practicing the "boy method."

I decided to give the "girl method" a try. Here, too, the evidence was convincing. Some friends who had girls were able to confirm that they were conceived under the circumstances outlined in the method. I even predicted that a relative's boy would be a girl because it was a "surprise," and indeed it was. After six long, and sometimes frustrating, months, I became pregnant. My cycle was irregular and hard to interpret at times. But persistence paid off, and . . . our daughter was born. It was one of the happiest days of my life.

D.S., Atlanta, Ga.

[From a completed questionnaire returned to us, dated May 20, 1987:] We are still on cloud nine! Thank you so much for making your information available to us. Our family is now complete with the birth of our son. . . . Looking back at our first two pregnancies, we are sure that your instructions for trying for a girl work equally as well. Thanks again!

Mrs. S.L., Vernon, Calif.

[From a completed questionnaire returned to us, dated December 1, 1987:] After five girls, thank you for our boy!

Mr. and Mrs. J.M., Portland, Ore.

March 11, 1988
Dear Dr. Shettles:

After being told by two doctors that we could not conceive at all, because of my husband's low sperm count, we followed your advice for the boy and our son was born last month—healthy as can be. I had two daughters by a previous marriage, so everybody is happy. My husband would have been thrilled fathering a child of either sex, after believing so long that he was basically sterile, but having a boy to go with our two daughters has exceeded his wildest hopes.

Mrs. J.A., Monroe, La.

September 27, 1989
Dear Dr. Shettles:

After my sister-in-law used your method to give her two daughters a son, I finally overcame the skepticism of my engineer husband and we tried for a little girl to go along with our two rambunctious boys. Even then, however, my husband made a bet with me it wasn't going to work. My husband was with me in the delivery room, and the first words out of his mouth were "I can't believe it." And the next words out of my mouth were "You owe me a hundred bucks." He paid up on the spot. Our little girl is six months old today.

A.Z., Pocatello, Ida.

May 17, 1990
Dear Dr. Shettles and Mr. Rorvik:

Three girls and a boy at last! Politically correct or not, we were going to keep trying till we got one. Thanks to you, our own little population explosion is over!

Mr. and Mrs. J., Provo, Utah

June 2, 1991
Dear Dr. Shettles:

Odd as it may sound we wanted only one child and we both wanted it to be a girl. I grew up in a family with nothing but brothers and so did my wife. We followed your method exactly, practicing through several months before trying the "real thing." Our daughter is three months old today and we couldn't be happier.

V.I., Chicago, Ill.

[From a completed questionnaire returned to us on June 9, 1992:] We followed all the recommendations and had our last intercourse four days before ovulation. But I was still expecting to have a boy because that's all they've had on my husband's side of the family. I had some pink undershirts and diaper covers still in their original packages with receipts so they could easily be returned. It was with great joy—and a great surprise—to find we had a girl!
Sincerely, J.J., Portland, Me.

January 12, 1993
Dear Dr. Shettles:

The first time we tried for a girl, we had a boy (I wanted a girl, then a boy). We used the book by Elizabeth Whelan. She says to have intercourse close to ovulation for a girl. As you know, your theory is opposite. Your latest edition did a good job of explaining why her book is not accurate. Your previous book did not mention it, probably because hers was not out yet? She tore apart your theories. Because you did not have an opportunity to defend your book, we believed her. I am so glad we purchased your book. I am so fulfilled now having a boy and a girl. I can't describe the feeling. Thank you.
H.B., Springfield, Ohio

March 27, 1994
Dear Dr. Shettles:

We had a great laugh—but it worked!
C.B. and N.S., Pomona, Calif.

November 1, 1994
Dear Dr. Shettles:

Thank you for your book. My husband and I are Christians, and I do believe that regardless of attempts suggested by

your book, the sex is predetermined by God, though I don't think it hurts to make His job easier.
Sincerely, Mrs. E., Roanoke, Va.

May 1, 1995
Dear Dr. Shettles:

Your approach was easily understood and not threatening. Without offering too much hope, you provided us with the confidence to try once more for the little girl we prayed for after already bringing into this world four handsome and very intelligent little lads. In April we gave birth to our daughter. Her four brothers will gladly protect and care for her. Thank you again.
Mr. and Mrs. W.V., Taos, N.M.

October 17, 1995
Dear Dr. Shettles and Mr. Rorvik:

Many years ago, in 1982, I got a book from the library that described how to try for a boy. I'm sure it was yours. I had a boy in 1982 and tried for another in 1985. Then my husband and I decided—with a lot of coaxing from our two sons—to try for a girl. We bought two books, yours and Whelan's. I trusted yours more because of our success with the boy method. Sure enough—we got our girl! Thanks so much!
Mrs. D.A., Los Angeles, Calif.

February 19, 1996
Dear Dr. Shettles:

I am writing this letter to express my deepest gratitude on the happiness that your book has bestowed on our family. I read an article in one of the parenting magazines, and it mentioned your book. I ran right out and bought a copy. We'd already had a son and, for financial reasons, did not feel we could have more than one more child. We wanted a girl to balance things out. We were aware that the possibility of a girl was lessened by the fact that my husband could not find a female in his father's line for at least three generations. Earlier this year I delivered a beautiful baby girl. She is a miracle and is loved by her big brother. Our family feels so deliriously happy that I have recommended your book to many friends of ours. Thank you for the insights into the wonders of conception and for the wonderful result we obtained.

T.O., Reading, Pa.

June 22, 1996
Dear Dr. Shettles:

In August, I found the mucus, temperature and LH surge all to coincide. We did everything you said in your book, trying for a boy, and last month our twin sons were born! Thank you! Mrs. J., Ft. Worth, Tex.

Since our last revision in 1996, we have continued to receive testimonial letters and questionnaires. In addition, several online Internet sites that sell our book have published testimonial letters, both pro and con. Major among these is Amazon.com, biggest of the online booksellers. We thought it would be interesting to compare the results they were reporting to our own. So, we selected, at random, a day in early 2006 to analyze *all* (negative as well as positive) of the letters published at that site related to our book. As we said earlier, no "cherry-picking," only the best of the crop.

This site had published more than 120 letters, dating back to early 1998 and continuing into 2006. We found it fascinating that in this sample of letters approximately 80 percent reported success and 20 percent reported failure—almost exactly what we have been reporting for the past several decades!

We invite you to look at these letters online for yourself—to see what still more couples have to say about the Shettles method and its competitors. You may also find some of these testimonials instructive in your own efforts and learn more about what to emphasize and what to avoid in your own sex-selection planning. You can, in addition, find more letters related to our book at the Barnes and Noble online site.

In addition, of course, we have received thousands of letters in the past ten years, since the last edition was published. Here is a sampling:

January 4, 2000
Dear Dr. Shettles:

What a way to start the new millennium! Our son was born at 5:30 A.M. on New Year's Day, joining our delighted daughters. And we do mean daughters. THREE of them to be exact! This is our last baby, and we couldn't be more thrilled. We did everything you recommended, and you delivered the happiest New Year of our life.
A.J.J., Chicago, Ill.

June 11, 2001
Dear Dr. Shettles:

We want to thank you for our daughter, but, more than that, we want to thank you for having a healthy baby. We believe that your method enabled us not only to have the daughter we wanted but to get pregnant in the first place. After two miscarriages and no successful past pregnancies, I had about given up until I came across your book. Both my husband and I found it easy to understand and follow and very sensible.
Mrs. L.V., Amman, Jordan

March 17, 2002
Dear Dr. Shettles and Mr. Rorvik:

We believe we followed your advice to the letter, but we did not succeed. This was a big disappointment since our neighbor had used your method twice with success. She is still convinced we failed to do everything you advised, but we are convinced otherwise. Well, we knew it wasn't foolproof, and we are happy to have a healthy new baby boy, who joins two brothers. No more children for us.

Mr. and Mrs. R.Y., Hot Springs, Ark.

July 8, 2003

Dear Dr. Shettles:

I had to convince my husband, who is a dentist, to try your method. He did not feel it was "scientific" enough, even though one of his doctor friends used it with success. He kept trying to find fault with it and warned me that we would end up with another boy—we have two already—if we tried it. He said that was fine with him, but he knew how much I wanted a daughter. I told him I'd take the chance. He is singing a different tune today. Our daughter was born two weeks ago. And guess who your newest booster is? One very pleased dentist. (I was already a booster.)

Mrs. A.A., Los Angeles, Calif.

December 7, 2004

Dear Dr. Shettles:

I have a sister who breeds dogs, and she claims (don't ask me how!) she has used it to select canine gender! She regards dogs as pretty much akin to humans, so this doesn't surprise me. I also know of two others who have used it with reported success in the human realm. My husband and I will use it next year to try to balance out our own family.

Mrs. A.U., Nashville, Tenn.

October 2005

Dear Dr. Shettles and Mr. Rorvik:

Thanks for writing a terrific book. It didn't work for us, but, looking back, we can see where we failed to time ovulation correctly on the critical try. We just got impatient and jumped the gun. We still recommend your book to all our friends, as two of my sisters—of the more methodical type—have used it successfully.

Mrs. B.P., Albany, N.Y.

March 17, 2006

Dear Dr. Shettles:

Wow! We succeeded on our first try. A baby girl! This is going to take some getting used to after three boys! Break out the pink paint! Thank you!

Mr. and Mrs. K.G., Omaha, Neb.

Not Everyone Succeeds

It's always a thrill for us, no matter how often it happens, when we hear from couples who use sex selection successfully. And it's always disappointing to us personally when we hear from couples who have tried and "failed." It's wise to remember that not *everyone* succeeds. A failure rate of up to 25 percent is not insubstantial. Some of these failures result from carelessness or failure to follow directions. Sometimes, however, couples appear to do "everything right" and still don't conceive a child of the sex they want.

Consider, for example, the woman we heard from in November 1979. She reported, via one of our questionnaires, that she and her husband had failed in their attempt to have a boy. "I do believe we carefully carried out all of the recommendations in your book," she wrote. We examined the information she gave us, including her monthly charts, and had to agree with her. This couple gave every indication of having done everything precisely "by the book." Moreover, the woman's cycles were not particularly irregular. The only unusual factor we could find was a preponderance of female offspring in the husband's family history.

Despite the fact that this couple did not get the son they wanted, they told us they now regarded their family as complete with two daughters. They added that they were happy their new baby was healthy and said, "She is a real joy to us!" This couple obviously approached sex selection with a realistic attitude and were prepared to be happy with a new baby, whatever its sex. We applaud them.

Sometimes the reasons for failure are obvious; many times those who report them are quick to point out the reasons for lack of success themselves. One woman wrote: "I have come to the conclusion that no one is at fault." She noted that her cycles are extremely irregular and that her cervical mucus is exceptionally scanty, even at the likely time of ovulation, making proper timing very difficult. She noted, however, that she was studying different ways of determining ovulation time in greater depth and, in fact, made some suggestions that we have found helpful. So we, too, learn from the "failures."

Those who meet with "failure" often do so with good humor. Recently, for example, we heard from a doctor who said he and his wife had tried our method for a girl, since they already had a boy. They did not get the daughter they wanted, and in a "P.S." the doctor added: "I had twin boys! *That* will teach me to tempt fate!"

In the many years that we have been receiving feedback on sex selection, we have had only one really irate letter. It came from a woman who said the method had failed for her not once but twice, and she now had two sons and no daughters. She said she was "very disappointed" and accused us of raising "false hopes." As we said at the outset of this book, and as we have always said, in fact, sex selection is not for everyone. We've put the word "failure" in quotation marks because we don't consider giving birth a "failure" just because the baby is not of the sex that might have been preferred. Those who do should not be using sex selection.

This particular letter made *us* a bit irate. We think the "false hope" accusation is particularly unfair. But, in fact, we would rather be "guilty" of giving people hope—even if it isn't hope that will be fulfilled 100 percent of the time—than of giving what we call "false despair." That, unfortunately, is what the majority of couples who go to their doctors with requests for sex-selection information receive. These couples are told, rather dogmatically, that sex selection is still a fifty-fifty affair, that "nature" decides these things and "there is nothing you can do about it." Nonsense. Read on.

Results to Date

We have reported over the years that our success rate is at least 75 percent. It has always been somewhat higher among those seeking boys than among those seeking girls. There are reasons for this that will become evident as you study this book. In the past five years we have further analyzed our results—in more demanding ways than we had used previously.

Our overall success rate for female conceptions is a tiny fraction under 75 percent. Our overall success rate for male conceptions is just a hair over 80 percent. With those who present good evidence—on the basis of their charts and their completed questionnaires—of following all instructions carefully, the success rates go up. But since to err is to be human, we believe it is best to "advertise" a success rate of 75 to 80 percent. Simply be aware that with extra care your individual chances may be considerably better. Moreover, with the advent of additional, sometimes more reliable ways of determining ovulation time—reported on in this edition—we expect our success rate to be even higher in the near future.

In the previous edition of our book we asked readers to notify us *in advance* as to whether they intended to use our method for the boy or for the girl. We hoped thereby to be able to follow up on more cases. Unfortunately, this has not proved practical on a large scale. We would require a large office staff—and some sizable funding—to be able to do this. Nonetheless, we have been able to match hundreds of advance notifications with subsequently submitted questionnaires. These kinds of data are considered more reliable than data that are submitted wholly "after the fact" or "retrospectively."

When we look at this smaller but still significant body of data, we find that the results run closely parallel to the sample of respondents as a whole. This is highly encouraging. Among those who notified us in advance that they intended to try for female offspring (and later sent us completed questionnaires with the results of their efforts), we find the success rate only slightly lower than that for the sample of girl-seeking couples as a whole—approximately 73 percent. Among those who notified us in advance that they intended to try for male offspring (and also later sent us completed questionnaires with their results), we obtain a success rate virtually the same as that enjoyed by the larger sample—80 percent.

We would like to thank those who participated in this study and regret that we were unable to contact each and every one of you.

Some Things We Would Like to Hear from You About

Before we move on we'd like to take this opportunity to ask *you* a few things we hope you might tell us about in the future. You can be thinking about these things as you proceed with the adventure of sex selection.

How many of you plan to tell your future child that he or she was the result of planned preconception sex selection? Why will you tell the child? Why won't you? How many of you will tell friends and relatives? Why will you do this or why won't you?

How many of you know other couples who have used our methods—or other methods of sex selection? What were the results of those efforts, so far as you know?

These are things you can write to us about when you send us your comments, report results, or ask questions of your own.

Now let's take a brief look at the history of sex selection and see how we arrived at the modern Shettles method.

Part Two

The Emergence
of a
Scientific
Sex-Selection
Method

Hundreds of Years
of Trial and Error
(Mostly Error)

It seems likely that human beings have been interested in the sex of their offspring almost from their first days on this planet. In fact, in some past eras there has been far more anxiety than there is today over the gender of offspring. There has, particularly, been a lot of anxiety over the production of male offspring—to prove the "virility" of men, perpetuate family names, work the land, fight in armies and clans, and so on. Great kingdoms have been thrown into chaos simply because a queen—or a whole series of them—did not produce a male heir. Countless numbers of wives, at all levels of society, have been cast out or even killed because they "failed" to produce the sons their husbands demanded.

We have no doubt that at least a few ancient practitioners of sex selection were also killed—when their conjurings and concoctions failed to produce the desired result. Those early practitioners of sex selection were really up against it. They were often damned if they did and damned if they didn't. If they tried to influence conceptions, there was often someone in authority standing by to accuse them of "meddling in the Lord's work," practicing witchcraft, or worse. Yet couples would continue to importune them, demanding their help. Then, if they failed, it wasn't the Lord's wrath they had to look out for but the wrath of a disappointed husband, distraught wife, and/or a whole family clan.

Our predecessors in this field could not have lived placid lives, particularly since there was still one more strike against them: knowledge of the human reproductive system was, until relatively recently, poor to nonexistent.

Right for the Boy, Left for the Girl

Even the ancient Greeks, with their scientific bent, were off the wall when it came to sex selection. Suppose you and your spouse had gone to consult Parmenides of Elea, a Greek savant of the fifth century B.C. If you had told Parmenides you wanted a boy, he would say (forgive our "translation"): "Okeydokey, now listen carefully—when you go to bed tonight, Helen, recline on your *right* side. Then when Achilles here has knowledge of you, his semen will flow into your *right* womb instead of your *left* womb. And *that* should do the trick, kids, because, as all we wise men know, boys are made in the right womb and girls in the left womb. Have a nice evening, and don't forget to pay my receptionist on the way out."

Well, there was at least *some* reason behind Parmenides' recommendations. He—and others of his era—had cut open various animals and discovered that they have *two* uteruses. Naturally, they assumed humans did, too. Why, they asked themselves, should there be two uteruses? Why not, they answered themselves, since there are two sexes? How they settled upon the right for the male and the left for the female is anybody's guess, though over the centuries there has been a male-chauvinistic tendency to ascribe all things right-handed (i.e., all things "just and good") to men and all things left-handed (i.e., all things "devious and sneaky") to women.

Another early-day sex selector, Anaxagoras, was also hung up on the right-left thing. He was considerably ahead of his time in at least one respect, though: he was convinced that it was the male and not the female who determined the sex of the child, something the science of our own era would confirm. Unfortunately, his ideas about how the male did this were nowhere near the mark. He decided that the products of the right testicle produce boys and those of the left, girls.

Those who attempted to select sex on the basis of these right-left theories almost certainly did a lot of thrashing about in bed, trying to come up with the winning configuration. There were many variations that could be tried, depending on who you were consulting at any given time. Some advisers insisted that both partners had to be on their right side during intercourse if a boy was to be achieved, or both on their left side if girls were wanted. Moreover, some insisted that, for the boy, the right ovary and the right testicle be in a prescribed alignment, which is easier said than done. A still knottier formula called for the male to tie a string tightly around whichever testicle corresponded to the sex that was *not* wanted. It was presumed that the string would put said testicle out of commission, a presumption that was possibly and painfully correct. Hippocrates himself, the "father of medicine," came up with that one.

Democritus and Aristotle were two other early-day sex selectors. They believed that males and females both produced semen. If the "female semen" predominated, then a girl would result; if the "male semen" prevailed, then a boy was in the works. Democritus, however, was also ahead of his time, at least in declaring that even though the sex of the child is determined by one parent, its other characteristics are the product of an intermingling of the male and female elements. Thus, though a boy would inherit his sexual identity from his father, he could still inherit his mother's eyes or smile. This "intermingling" theory foreshadowed our modern knowledge of genetic heritability.

Aristotle took a more "activist" stance. It wasn't enough for him to tell people that one "semen" predominates over the other. There was another matter to be considered: "vigor." You could *make* your semen the more powerful if you would simply throw yourself into the procreative act with "vigor" greater than that of your partner. Of course, there were some things that affect "vigor" that might occasionally be beyond your control—such things as the weather and the way the wind blows. These factors could be used to advantage, however, by careful timing of intercourse.

"More males are born if copulation takes place when a north wind [rather] than when a south wind blows," Aristotle wrote, "for the south wind is moister. Shepherds say that it makes a difference not only if copulation takes place during a north or a south wind, but even if the animals while copulating look toward the south or the north. So small a thing will sometimes turn the scale."

God help the ancient Greeks—who not only had to throw themselves, vigorously, if at all possible, right and left but also north and south!

Lion's Blood, Nut Trees, and Raw Meat

By the middle Ages, things had not improved. In fact, they had worsened. Whereas the Greeks had at least made certain observations the basis of their sex-selection theories, the alchemists of the Middle Ages relied on superstition. If you wanted a boy during those dark days, you would be advised, according to some texts, to obtain a mixture of wine and lion's blood from your neighborhood alchemist. You'd drink this just before copulating under the full moon. It was necessary, in addition, to have an abbot praying nearby, on your behalf, though whether he was praying that you would beget a boy or that you would not catch your death of cold out there under the full moon or die of lion's-blood poisoning is not clear.

Coming closer to our time, superstition has continued to play a role in the sex-selection beliefs of some. Young brides of the Palau Islands dressed up in men's clothing before intercourse, thinking this would help them conceive sons. In Sweden, for a time, girls took young boys into their beds on the eve of their weddings, thinking this would similarly help them conceive male offspring—the following night. In parts of Yugoslavia the no-doubt-bewildered boy spent the wedding night in bed with both the bride and the groom, while the couple tried to conceive a boy of their own.

In parts of Germany, woodsmen who wanted sons used to take axes to bed with them. Then, while having intercourse with their wives, they'd chant, "Ruck, ruck, roy, you shall have a boy!" If a girl was desired, the woodsman would dispense with the manly ax and climb into bed undefended, chanting, during intercourse, "Ruck, ruck, rade, you shall have a maid!"

Some of these folk methods are said to be in use in some rural and remote parts of the world even today. A few men in the backwoods of the United States are said to still be hanging their pants on the right side of the bed if they want a boy and on the left side if they want a girl. In some Slavic countries a few wives apparently still pinch their husbands' right testicles, though whether to stimulate or discourage male conceptions we're not sure. Perhaps it's merely to discourage conceptions of *either* type. In parts of Italy, meanwhile, it's the ears that take a beating. The man bites his wife's right ear if he wants a son, her left ear if he wants a daughter.

In Austria there are a few who may still believe, as many there once did, that a year with a good nut harvest will also be a good year for having sons. The "reasoning" behind that is obvious enough. Things can be helped along even more, though, if a midwife buries some afterbirth under a nut tree. That will help ensure that the next baby is a boy—or so it's claimed.

In many parts of the world there are couples who believe that the sex of the child is determined by the "stronger" or "wiser" of the two marital partners—a latter-day expression of some of the Greek ideas. Some think the child will be the same sex as the older of the two partners. Others think gender is something that is not determined until well into pregnancy and can thus be influenced over a considerable period of time. One notion is that if a woman persistently eats sweets during her pregnancy, the child will "turn out" to be a girl; if, on the other hand, the mother tears into a lot of good, red meat, the baby will, of course, be a boy—and perhaps a brave and snarling one at that.

Mind Over Matter—and Gender, Too

Among the most intriguing sex-selection theories was that of C. Wilbur Taber, as put forward in his imaginative 1899 book, *Suggestion: The Secret of Sex*. It was Taber's idea that the power of suggestion alone could determine which sex would result. He recommended that husbands practice a technique that amounted almost to hypnosis of their wives, putting them nearly to sleep while simultaneously implanting the idea that a boy would be conceived. This "suggestion" was to be followed by the "procreative act," preferably with as little participation from the ideally stuporous woman as possible. Later on, if the woman didn't nod off altogether, Taber advised the husband to read to his wife from the biographies of great and famous men.

Others eagerly followed up on Taber's work. W. Wallace Hoffman, M.D., published a book in 1916 called (are you ready?) *Sterility and Choice of Sex in the Human Family: A Subject with Which Medical Books Do Not Deal, Having Special Reference to the Causes of Change in and the Theories of Determination of Sex in the Unborn Together with a Little Thought on the Question of Sterility.* Dr. Hoffman believed that one's thoughts could determine not only whether one would conceive a boy or a girl but also whether said offspring would be, for example, a drunk or an idiot. One had to be cautious about one's frame of mind while attempting to conceive. Think the "wrong" thoughts for just an ill-timed moment and little Bobby might grow up to rob banks or little Mary might mature into a lady of the night.

Cary S. Cox, whose book *Causes and Control of Sex* appeared in 1923, went a step further: yes, you could choose the sex of your child with the right thoughts, but be careful or you might give birth to one who resembled *a turtle!* Cox claimed this had happened to a woman who had been badly frightened by a turtle while pregnant and subsequently gave birth to a baby with turtlelike appendages.

Numerous books have appeared on the effects diet can have on gender. (We'll have more to say about this in a later chapter.) Dr. Frank Kraft, in his book *Sex of Offspring: A Modern Discovery of Primeval Law* (1908), insisted that optimal nutrition would result in female offspring. John McElrath, on the other hand, in his book *The Key to Sex Control or the Cellular Determination of Sex and the Physiological Laws Which Govern Its Control* (1911), said just the opposite: good diet would beget boys.

A good many books on sex selection in the twentieth century have harked back to the Greeks. The more "vigorous" or more "noble" marital partner, it has been claimed in many of these tomes, determines the sex. Still, if you were the "weaker" partner, you might yet have a say in things, provided you were clever. Some authors suggested that if you had a particularly strong wife, for example, you could cheat a little by waiting for her to become ill before trying to conceive a child. Or, if you were a ninety-eight-pound weakling, you could always lift weights. If you were more clever still, you might engage your superior but less devious partner in psychological warfare, berating and gradually wearing down said partner, stripping him or her of some of that strength or nobility. If that didn't work, there was always outright violence to rescue you. One author went so far as to propose rape as a means of diminishing "a woman's superiority"!

Perhaps the most avid believer, in the twentieth century, in the old right-left theories of the Greeks was E. Rumley Dawson, a fellow of the Royal Society of Medicine. He argued that it was the woman who determined the sex of the child—and the woman alone. Eggs from the right ovary produced boys, he claimed, while those from the left ovary produced nothing but girls. In numerous magazine articles and in his 1917 book, *The Causation of Sex in Man,* Dr. Dawson suggested ways to take advantage of this "knowledge." It was his idea that the ovaries take turns ovulating each month. Couples who followed Dawson's advice thus spoke of "little-boy months" and "little-girl months."

The system was simplicity itself, for after the birth of a first child (the sex of which could not be planned), you merely paid attention to what month it was and timed your next attempt according to the month/sex desired. What explanation was given when couples failed to produce children of the expected sex we do not know. Perhaps they were merely told they couldn't count.

While some paid attention to left and right, others were heeding the phases of the moon and the positions of the stars. A number of sex-selection theoreticians have plotted *astrological* means of selecting the sex of children. As recently as 1973, in fact, a book appeared on this subject: *Natural Birth Control and How to Choose the Sex of Your Child* by Lynn Schroeder and Sheila Ostrander.

Farmer Rummins and His Heifers

One thing that should be kept in mind while reviewing the "old," the "folk," and the "nonscientific" methods of sex selection is that some of them may actually have had some validity, at least in part. We are reminded, in this context, of a delightful story the noted British author Roald Dahl once related in the pages of the *New York Times.* Dahl claimed the story is true. It involves a "Farmer Rummins" and his magnificent herd of dairy cattle. Dahl, his mother, and his sister lived near the Rummins farm and, for a time, kept a cow themselves. When they decided to breed the cow, they took it to be serviced by Farmer Rummins' prize bull. The farmer asked whether they wanted a heifer or a bull calf. Young Dahl said that, given a choice, he'd prefer a heifer, since it was more milk and not beef that the family needed.

No problem, said the farmer, whereupon he prodded the cow until it was facing squarely into the sun. Only then did he permit his bull to mount it, during which time he took care to see that the cow *kept* facing into the sun. All of this, the farmer insisted, would ensure that the Dahls got the heifer they wanted. When Dahl expressed doubts, Farmer Rummins bristled. "Don't be so damn silly," he fumed. "Facts is facts."

The farmer then trotted his young neighbor into the house and sat him down in front of a stack of old ledgers in which were recorded each and every bovine mating that had taken place on the farm—dating back some thirty-two years. There were columns in the ledgers for the date of mating, date of birth, and the sex of the calf. This was a dairy farm and so, of course, heifers were what Farmer Rummins had wanted—and heifers were what he had got. As Dahl's finger flew up and down the long ledger columns, there was, under the sex-of-calf heading, a monotonous repetition of the word "heifer."

Next to one of the few bull-calf entries, Dahl found this notation: "Cow jumped around." Rummins explained that some cows just couldn't be kept facing into the sun during their brief encounters with the impatient bull. And when cows faced away from the sun during mating, they invariably gave birth to bulls—so said Farmer Rummins. Dahl was impressed enough so that he went through all the ledgers and counted up the bulls and heifers. The final tally was 2,516 heifers and 56 bulls. Why, asked an amazed young man, had the farmer kept this to himself for so long? "I reckon it ain't nobody else's business," Rummins answered. The procedure had worked for him and made him prosperous in the process—that was enough.

Rummins' father had been a dairy farmer before him and had passed the information on to him. "He explained to me," Rummins recalled, "that a cow don't have nothing to do with deciding the sex of the calf. All a cow's got is an egg. It's the bull decides what the sex is going to be. The sperm of the bull. According to my old dad, a bull has two different kind of sperm, female sperm and male sperm. . . . So when the old bull shoots off his sperm into the cow, a sort of swimming race takes place between the male and female sperm to see which one can reach the egg first. If the female sperm wins, you get a heifer."

The farmer's father was correct in this—and considerably ahead of his time. It is absolutely true that the male is responsible for determining the sex of the offspring, that the male produces two types of sperm, and even that the two types do, in a sense, "race" to see which will reach the egg first.

As for the rest of the farmer's theory, we can't vouch for that. It was the farmer's idea that the sun exerts an attraction, or "pull," something like the force the moon exerts on ocean tides, somehow giving the female-producing sperm an advantage. "And if you turn the cow around the other way," the farmer said, "it's pulling them backwards and the male sperm wins instead."

Dahl asked Rummins if he thought this method might work for humans, too. The farmer said it would: "Just so long as you remember everything's got to be pointed in the right direction. A cow ain't lying down, you know. It's standing on all fours. And it ain't no good doing it at night either, because at night the sun is shielded behind the earth and it can't influence anything." Did Rummins have any *proof* that the method would work in humans? The old farmer, Dahl said, responded with "another of his long sly broken-toothed grins" and a question of his own: "I've got four boys of my own, ain't I? Ruddy girls ain't no use to me around here. Boys is what you want on a farm."

Whether there is more bull to this story than meets the eye, we don't know. Sometimes it occurs to us that either the sly old farmer or the sly old storyteller might have wanted to see how many people would make jackasses out of themselves copulating on all fours outside in broad daylight while squinting into the sun. On the other hand . . .

The "Facts of Life" Revisited (What You *Must* Know Before You Attempt Sex Selection)

Before you try to select sex through "human intervention," it is essential that you understand exactly how Mother Nature selects sex. There can be no successful sex-selection method without a good understanding of the way in which sexual identity is conferred in the natural course of events. Old methods failed because they were based almost entirely on grossly imperfect knowledge, guesswork, and sometimes superstition. Let's review what scientists now know about the events that result in boys and girls. Much of it is fascinating; some of it is nothing short of amazing.

"X" Marks the Girl; "Y" Marks the Boy

As noted in the preceding chapter, it has been suspected for a long time that the seminal fluid of the male plays a crucial role in both the fertilization of the egg and the determination of the sex of the offspring. It wasn't until the seventeenth century, however, that anyone actually identified the sperm cells in the seminal fluid. And even then nobody knew for sure what they were or what they did. Some thought each cell contained a fully formed though exceedingly tiny human baby, which, once deposited in the womb, would grow into a larger human baby.

There continued for some time to be no *proof* that either semen or sperm was required for the reproductive process. Experiments with frogs in the eighteenth century finally demonstrated that some part of the seminal fluid was indeed necessary if fertilization was to occur. Eggs that were prevented from coming in contact with semen simply didn't produce tadpoles, whereas those eggs that were exposed to semen *did*.

Karl von Baer's work in the 1820s shed further light on the process. Von Baer dissected female dogs in various stages of pregnancy. He found embryos (fertilized eggs) attached to the lining of the womb and later discovered that, earlier in pregnancy, the embryos could be found in the fallopian tubes. He wondered for a while if these "ovules" weren't wholly the product of the male; he thought they might have been ejaculated, whole, into the female during intercourse. But then, following the tubes up to the ovaries of the female dogs, he discovered, in some of them, bulging follicles on the surface of the ovaries. And inside these follicles were eggs that looked much like the "ovules" he had been studying.

More of the puzzle began to fall into place a couple of decades later when other researchers, also working with animals, observed the actual penetration of the egg by the sperm. By the 1870s it was accepted that this combination of sex cells was required for fertilization and reproduction and that the male and female contributed about equally to the genetic inheritance of the offspring. In 1883 Pierre van Beneden showed more conclusively than anyone ever had that this was the case when he demonstrated that the sex cells have only *half* the number of chromosomes that the body cells (such as those that make up skin, bone, muscle, etc.) contain. Thus you had to get two sex cells to merge in order to produce a new cell with enough chromosomes to direct the development of a new human being.

Chromosomes, in case you've forgotten, are the microscopic units within cell nuclei that contain the even smaller genes, which, in turn, contain the genetic codes that determine our physical characteristics (eye and hair and skin coloring, etc.). It is through the combination of male and female sex cells—sperm and egg—that a new "master cell," or embryo, is created which contains the full set of chromosomes needed to produce a new human being, with a combination of genes unique to that individual.

Van Beneden's discovery, therefore, was an important one. But it still didn't tell us how nature determines the sex of the child that results from the fertilization of an egg by sperm. Nor did it tell us which of the two parents was responsible for this determination. It wasn't long, though, before an answer began to be suggested. Studies of the chromosomes in egg cells that had not yet undergone fertilization showed that the chromosomes were arranged, within the nuclei of the eggs, in pairs and that all of the pairs seemed to be perfectly matched. This was *not* true in the sperm cells, where one chromosome pair, in each cell, was unmatched. One of the chromosomes in this odd pair was noticeably smaller than its partner.

Researchers were not slow to seize on this "discrepancy" as the possible key to sex selection. The American zoologist C. E. McClung was the first, in 1902, to call this mismatched pair "the sex chromosomes." The smaller member of the pair was referred to as the "Y" chromosome and the larger as the "X" chromosome. When the sperm cells matured in preparation for the fertilizing task, they divided in half, with the chromosome pairs splitting up to go into new cells. It was hypothesized that the Y chromosome went into one of the newly formed sperm cells and the X into another. The guess was that eggs fertilized by Y sperm would yield offspring of one sex and the eggs fertilized by X sperm would yield offspring of the other sex.

This was confirmed in 1905—but only in the lowly mealworm. Dr. N. M. Stevens found that mealworm ova fertilized by Y sperm *all* developed into *males,* while those eggs fertilized by X sperm *all* became *females.* It took another twenty years to show that this pattern holds true for a number of more advanced animals, as well. No one had yet proved, however, that this was the pattern for humans. Human chromosomes, in contrast to those of many primitive animals, are so small that it wasn't until 1956 that their number was firmly established at forty-six for each body cell and at twenty-three for each sex cell.

In the 1950s Dr. John Rock of Harvard and Dr. Landrum Shettles of Columbia became the first researchers to watch the complete process of human conception under their microscopes. Through various laboratory techniques, the sperm bearing the X chromosome could be shown, at last, to produce girls and the sperm carrying the Y chromosome could be shown to produce boys.

The "Queen Mother of Cells"

Let's look now at how egg and sperm come together and how sexual identity is conferred in the process. We begin with the human egg, or ovum, which Dr. Shettles has called the "queen mother of cells." This is an apt label, for there *is* something regal about the ovum. Like the queen ant or queen bee, the queen cell is a giantess, at least when compared with other cells. It weighs a staggering ½0th of a millionth of an ounce and measures about ½75th of an inch in diameter. This makes it large enough to be seen by the naked eye, provided that eye is in optimal condition. The egg is queenly in another respect, too. It moves at a serene pace and is attended by a retinue of "ladies-in-waiting," a gathering of lesser cells, each of which was once a candidate to become a queen, an egg, but was "passed over" for one reason or another. These "nurse cells," as they are sometimes called, cluster around the queen, protecting and nourishing her as she makes her way from the ovary into and down the fallopian tube.

A girl, at birth, already possesses more than a million primitive egg cells in her ovaries. By the time she matures, there will be only a few hundred thousand of these left, the others having died off over the years. Among the survivors, only a few hundred will have any real chance of fulfilling their destinies. Candidates for fertilization don't begin emerging until after puberty, when the pituitary gland in the brain excretes hormones that set in motion a complex series of biochemical events. It is under the guidance of these hormones that the girl begins to mature into a woman and the female reproductive, or menstrual, cycle commences.

The *average* menstrual cycle is twenty-eight days in length. Ovulation—emergence of the egg from the ovary—occurs about midway through the cycle—on *average* at day fourteen. The egg erupts from the follicle on the surface of the ovary and, surrounded by some five thousand of its nurse cells, is picked up by hairlike projections of the fallopian tube and then swept into the tube, through which it must pass on its journey down to the womb. On its way down the tube, the egg sheds some of its shimmering "corona radiata," that surrounding halo comprised of clinging nurse cells. It begins to "undress," in effect, hoping for a rendezvous with a sperm cell somewhere in the tube.

If the queen is "stood up"—that is, no suitors arrive to court her—all is lost. An egg cell that came into being decades earlier and has "waited" all its life for this moment only to be disappointed in love will proceed to the womb and perish there. The same hormonal processes that stimulated the ovary and caused ovulation also prepare the lining of the womb for the possible arrival of a fertilized egg by thickening it and making it more receptive. But if the egg arrives unfertilized, the "bed" is stripped away. The succulent cells that have lined the womb, ready to nourish new life, are expelled via menstruation, or what the woman calls her "period." Then the process starts all over again with the next monthly cycle. Perhaps the next would-be queen will be luckier.

Millions of Sperm in Pursuit of One Egg

The sperm cells are quite different from the ova. Sperm, too, however, already exist in great numbers at the time the male is born. These primitive sperm cells are dormant within the tubules of the testes, the male sex glands, until puberty, when, again under the direction of complex hormonal instructions, they begin to multiply—at the rate of about 300 million per day. A single male ejaculate may contain as many as one billion spermatozoa. Quite an army, considering the fact that they will all be out to fertilize one egg! And since only one sperm usually penetrates the egg, this looks, at first blush, like massive "overkill."

In fact, however, the tiny sperm cells have got their work cut out for them. Only a very few will ever get near the egg. Sperm weigh only $\frac{1}{90,000}$th as much as the egg, and though they are longer than eggs are wide (the typical sperm is about ⅟₅₀₀th of an inch in length), most of that length is in the whiplike tail. These tiny cells have to get to the egg through a formidable series of obstacle courses. Most of them die along the way. It is only sheer numbers that makes fertilization a possibility at all.

Sperm cells are the smallest cells in the body. It has been estimated that all of the eggs that were required to produce the entire population of the world today could be contained in a single cookie jar. By contrast, you could put all of the sperm required for the same job into a thimble.

The troubles these diminutive soldiers face crop up the instant they are ejaculated into the female reproductive tract. They don't go serenely into the "outside" world cushioned and encased by protective nurse cells, à la the egg. Instead, as Dr. Shettles is fond of saying, they are booted out "stripped to the shorts." They are designed for speed more than for staying power. Dr. Shettles has also likened them to salmon that must swim upstream for many miles against a strong current. Make that an *acidic* current. The vaginal environment tends to be acidic, though it becomes somewhat more alkaline—and thus a bit more receptive—the closer one gets to ovulation, nature's way of giving the sperm a helping hand. Even then, though, the secretions can be hostile; a great many of the sperm in any given ejaculate die in the vagina fairly quickly.

Sperm that are "booted" all the way to the back of the vagina often have a better chance of surviving and getting to the egg. The secretions nearest the opening of the womb (the cervix) tend to be more alkaline. Still, even the sperm deposited near the cervis have to beat their tails furiously to get through the sticky swirl of fluid that is issuing out of the cervix. Those that do get inside the womb are still fighting adverse currents as they struggle toward the opening of the fallopian tubes. Then they have to make a choice. Which tube should they choose? The wrong choice will prove fatal for sure.

The sperm cells have no "radar" or other magical means of knowing which tube harbors the egg. Very recently some new research has suggested that the egg may emit what amounts to a kind of "molecular perfume" that the sperm can "smell." (The implications of this for sex selection and contraception are discussed in the Afterword of this book.) Many go the wrong way. Those that pick the right channel are far from home free. The current is still against them. This is the same current—consisting of muscular contractions of the tubal walls, the downward streaming of ovarian and other fluids, and the ciliary beating of hairlike cells that line the walls of the tubes—that helps propel the egg downward toward the womb. The tubes are not smooth conduits; their walls have countless little recesses and dead ends that trap many of the sperm. Then there are the female white blood cells; these fierce scavengers may swoop down at any moment and literally gobble up the sperm cells, which are treated like any other foreign "invader." It's a little like a video game. The hero—the sperm cell—has to get to the safety of the egg while fighting raging currents and acids, traversing sticky whirlpools, avoiding dead ends, and dodging killer cells. And the "clock" is running all the time. Neither the egg nor the sperm cells can live for long.

Only a few hundred sperm, or a few thousand at most, actually make contact with the egg. Some charge right on by, missing the giant by a heartbreaking hair. The fact that so few get to the egg may be part of nature's design. All of those obstacles we've spoken of may be a means of separating the men from the boys, so to speak, so that only the fittest survive and have any chance of becoming the queen's consort. The fleetest of the sperm may get to the egg in only a few hours. Of course, if the egg hasn't been released from the ovary yet, the sperm may have made their harrowing journey in vain. Some will be strong enough to "wait around" in hopes that the egg will still show up, at which point fertilization may occur. A few particularly hardy sperm may survive for up to four days, though most will not.

"The Dance of Love"

Under ideal circumstances egg and sperm meet, head-on, somewhere in the upper portion of the fallopian tube. The sperm cells literally bore into the egg; having come this far, they do not wait for an invitation.

The egg, as it turns out, has a tough hide, and so the sperm have to whip their tails vigorously, hoping to be the first to penetrate the outer layers of the egg and thus gain exclusive admittance into the egg's inner sanctum. And we do mean exclusive, because once *one* of the sperm cells passes beyond a certain point within the egg it is as if the queen says, "That's my man!" and then flips a switch that locks out all other would-be kings. The "front runner," who has more than a foot in the door at this point, keeps on going, penetrating ever more deeply into the egg. The others come up against an invisible wall and, try though they might, can't go any farther.

The sight of all of this—under the microscope—is quite remarkable. As the winning sperm moves on inside, proceeding toward the target—the nucleus of the egg—the other hundreds or thousands of sperm continue to whip their tails—*synchronously.* This creates an undulating effect, looking rather like tall grass dancing in the wind. And the effect of this phenomenon is to make the egg rotate with increasing speed and always in a clockwise direction. Dr. Shettles calls this spectacle "the dance of love." The unsuccessful suitors go on whipping things up until they die of exhaustion. Meanwhile their frantic activity seems to serve a purpose, helping to propel the egg, now in the process of merging with the single sperm cell that penetrated the interior, at an even faster clip down the tube toward the eagerly awaiting womb.

The single sperm that gets all the way inside is declared the winner, in effect, when it gets through the clear outer layer, or cytoplasm, of the egg. Once it gets beyond that point, biochemical events are set in motion that keep the other sperm from going any farther. A violent vibration occurs within the cytoplasm—almost as if the egg is having an orgasm. The sperm penetration also stimulates the egg to undergo its final maturation process. The sperm, on the way to the center of its new universe, sheds its "nose cone," the apparatus that helped it get inside in the first place, and also its tail, which has now served its purpose admirably.

All that is ultimately left of the sperm is its "pronucleus," its package of twenty-three chromosomes. The sperm pronucleus approaches the egg pronucleus, another package of twenty-three chromosomes, now at a respectful pace. Slowly the two pronuclei make contact; their membranes begin to dissolve at the point of contact, and the marriage of chromosomes commences. First they intermingle, then they form into pairs. Together, these forty-six chromosomes form the nucleus of a new cell and, in fact, a new human life.

The sex of that new human life is determined entirely by the type of sperm that wins the race and merges with the egg. Twenty-two of the sperm chromosomes match up with twenty-two of the egg's chromosomes to determine all bodily characteristics—except for sex. The one remaining sperm chromosome and the one remaining egg chromosome pair up to determine sex. The egg *always* contributes an X sex chromosome to this pair. Sperm, however, may contribute either X or Y sex chromosomes. If the winning sperm carries an X, then the resulting pair will have to be XX, which, to the geneticist, spells *girl*. If, on the other hand, the winning sperm carries a Y, the result will be XY, which means *boy*.

So much for Mother Nature's method. Now let's move on to the Shettles method.

The Shettles Method:
How It Developed and
Has Been Refined

Almost from the beginning of his career as a specialist in obstetrics and gynecology, Dr. Shettles was confronted by parents eager to choose the sex of their children. Up until the early 1960s, Dr. Shettles did not believe there was anything he or anyone else could do to beat Mother Nature's odds. "It looked like it was going to have to stay roughly fifty-fifty," he recalls. Still, in part because his patients kept prodding him, he began looking for "anything that might be exploited toward this end, particularly anything related to differences in the two sperm types we then knew existed."

The Two Sperm Types

The two types of sperm, as discussed in the preceding chapter, are the Y-bearing (boy-producing) sperm and the X-bearing (female-producing) sperm. It was known some years ago that the Y chromosome is markedly smaller than the X chromosome. Dr. Shettles hoped that this difference might be expressed in the overall head (nucleus) sizes of the two different types of sperm. This might provide a means of fairly easily identifying and perhaps of separating the two types for use in artificial insemination, and so on. Unfortunately, using standard microscopes and staining techniques, Dr. Shettles failed to discern anything that would *visually* distinguish the two types of sperm.

But he kept trying different approaches and, "one night," he recalls, "I decided to examine some *living* sperm cells under a phase-contrast microscope." This type of microscope, relatively new at that time, illuminates microscopic objects in startling ways, immersing them in halos of light that sometimes reveal hitherto unseen details. In expert hands, phase-contrast microscopy can produce images of greater depth, images that are closer to reality than those seen under some other microscopes. The phase scope lent an electrifying new dimension to the living sperm, which darted across the field of vision looking a little, Dr. Shettles says, "like electric eels." He decided he could have a still better look if he could just slow down the sperm somehow. He did this by introducing a small amount of carbon dioxide gas into the fluid through which the sperm were swimming. The gas put the sperm cells into "slow motion."

"The very moment they slowed down I could see the difference," Dr. Shettles asserts. "There were two distinct populations. I was so excited that I ran upstairs and grabbed the first lab technician I could find. I had to show somebody what I'd found."

Dr. Shettles promptly published his findings in the May 21, 1960, issue of the prestigious British journal *Nature*. On June 5, 1960, this discovery was reported in the *New York Times*.

Dr. Shettles continued to examine sperm specimens, and his observations continued to hold up. In most specimens he'd find more of the round-headed sperm than the larger, oval-headed type. This finding was in keeping with the established fact that far more boys than girls are conceived (various studies have shown 110 to 170 male conceptions for every 90 to 100 female conceptions) and that for every 100 female births, there are approximately 105 male births.

Occasionally Dr. Shettles would come across a semen specimen that contained sperm of unusual uniformity, being almost all of one or the other type. In cases where these were of the larger, oval shape, Dr. Shettles recalls, "I'd find men who had given birth mainly to girls." And where the uniformity was of the other type, "I'd find men who had fathered an excessive number of boys."

Early in his studies, Dr. Shettles encountered a specimen that contained *nothing* but small, round-headed sperm. The sperm donor, Dr. Shettles discovered, came from a family which, for 256 years, had produced almost nothing but male offspring. During two and a half centuries, only *two girls* had been born to the men in this family.

Speed Versus Staying Power

From these early studies, Dr. Shettles reached three conclusions that were to be of great importance in the development of his sex-selection theories: (1) the male-producing sperm are smaller and more compact than the female-producing variety of sperm; (2) the male-producing sperm are, therefore, very likely *faster* than their bulkier, female-producing counterparts; (3) the female-producing sperm, on the other hand, because of their greater bulk, probably have *greater staying power.*

Even though there is still a dispute over whether it is possible to *visually* distinguish between the two types of sperm, what is important is that others *have* confirmed beyond any doubt what Dr. Shettles hypothesized: that the Y-bearing, male-producing sperm are, indeed, smaller and faster than the X-bearing variety. This has been demonstrated by means other than phase-contrast microscopy, as will be discussed in more detail later in this book. Dr. Shettles and some others still believe the differences *can* be visualized under the phase-contrast scope, given the proper preparation and requisite expertise. But, for all practical purposes, that issue is moot. It has been rendered beside the point by other—more persuasive—confirmations of Dr. Shettles' conclusions with respect to sperm speed and size.

By 1960 Dr. Shettles had a starting point, a place at which he could begin building his sex-selection theory. It appeared to him, as already noted, that there were far more male-producing sperm than female-producing sperm in the average ejaculate. The numerical advantage of the Y sperm, he reasoned, must be a sort of compensation for their fragility, relative to the huskier X sperm. The male, it seems, starts out life, even at the preconception stage, in a state of "inferiority," a state that is borne out all through life and is evidenced in the fact that there are far more stillbirths and miscarriages of male children than of female, in the fact that more boys than girls die in infancy, in the fact that women have longer life spans and are more resistant to many types of disease, and so on. It is only by having a numerical advantage at the outset that the male keeps roughly even with the female.

That advantage is taxed, Dr. Shettles hypothesized, from the very beginning—when the sperm are ejaculated into the vagina during intercourse. To test this hypothesis, he placed sperm in transparent tubes filled with cervical and vaginal fluids of varying degrees of acidity and alkalinity, such as naturally exist in the female reproductive tract. He found that when the fluids were most acidic, the larger, oval-shaped sperm were able to survive longer than the smaller, round types. He also found that when the fluids were highly alkaline, it was the male-producing sperm that would most quickly swim to the opposite ends of the tubes, which he likened to "physiological racetracks." When the secretions were more acidic, the likely "winners" were the female sperm. The acids took a toll on both types, but the toll was heaviest on the more fragile boy sperm.

Now, in the normal course of events, a woman's secretions tend to become most alkaline as ovulation approaches—nature's way of enhancing chances for fertilization. As you move away from ovulation in either direction, either before or after, the secretions become more acidic. Experiments showed that the female-producing sperm could not only better withstand the acids, they could also better withstand a number of forms of stress, such as heat, toxic chemicals, and so on.

The Importance of Timing

With this information in hand, it became evident that *timing of intercourse relative to the time of ovulation* might be the most crucial element in sex selection. Dr. Shettles theorized that intercourse that occurs at or near the time of ovulation will, more likely than not, result in *male* offspring. This is because the secretions in the reproductive tract of the female are most alkaline at or near ovulation and thus most favorable to the speedy Y sperm. But if the intercourse that results in conception occurs two or three days before ovulation, when the secretions are more acidic, then the chances for female offspring would be enhanced. This is because the hardy X sperm are more likely to survive in the acidic environment during the lengthy wait for the egg. This is where the X sperm's superior "staying power" comes into play.

Dr. Shettles sifted through the scientific and medical literature to see if he could find further support for his blossoming theory. He found that some past practitioners of sex selection had made observations with respect to acidity and alkalinity. Dr. Felix Unterberger, a German researcher, had, decades earlier, in the 1930s, treated some forms of infertility with baking soda douches. He had noted that a number of his infertile female patients had extremely acidic secretions. He guessed that the acids were adversely affecting the sperm. When the women used the douches containing the alkaline baking soda, the acids were neutralized somewhat, and a number of these women became pregnant. Equally astonishing was the fact that so many boys were born to these women—far more, Dr. Unterberger and others reported, than would have been expected by chance. Some other doctors followed up on the German researcher's work with similar results.* Even Emperor Hirohito of Japan is said to have employed an alkaline douche in his successful quest for a male heir.

Dr. Shettles found neither truly adequate confirmation nor refutation of Dr. Unterberger's work in the literature. Unterberger's theory had enjoyed a brief popularity and then had faded away for lack of rigorous scientific follow-up by other doctors, most of whom seemed to find the notion of alkaline douches beneath their dignity. Nonetheless, the data from the 1930s was intriguing and certainly tended to support Dr. Shettles' own developing ideas about sex selection. He doubted that alkaline douches alone would produce boys with any great reliability. Timing appeared to him to be a far more important component, but douching might be a potentially useful *adjunct* to timing.

---

*Reported in the medical journal *Lancet* and elsewhere.

Dr. Shettles also found numerous references, dating back to the nineteenth century, to the alleged influence female orgasm can exert on sex selection. Some researchers have claimed that when a woman has an orgasm during intercourse and, especially, when she has her orgasm before that of the man, the chances for having a boy are increased. If there was any truth to this persistent claim, Dr. Shettles decided, it might have to do with the fact that female orgasm makes the secretions somewhat more alkaline.

Even the Talmud, compiled centuries ago, notes a connection between female orgasm and the sex of offspring. "The determination of sex takes place at the moment of cohabitation," the Talmud declares. "When the woman emits her semen before the man [when she has orgasm first], the child will be a boy. Otherwise it will be a girl." The man, if he wanted a boy, was directed therefore to "hold back" until after his wife had an orgasm.

This was interesting in and of itself, meshing as it did with the reports of researchers in more recent times. But Dr. Shettles found another directive in the Talmud that was equally or even more interesting. Orthodox Jews were instructed not to engage in sexual intercourse during a woman's "unclean" period, meaning during her menstrual period; nor should they have intercourse for a full week thereafter, declared the Talmud. Orthodox Jewish authorities Dr. Shettles consulted confirmed that many Jewish couples abstain from intercourse for the first two weeks of the cycle; thus the first intercourse of each cycle would occur at or near the time of ovulation for most of these women. Taking this into account and considering the less important but still noteworthy directive on female orgasm (which might be expected to increase alkalinity), Dr. Shettles wondered if he hadn't found the reason why Orthodox Jews have a disproportionate number of male offspring, as has been reported for decades.

Of still greater importance to Dr. Shettles was the data on artificial insemination. It seemed clear to Dr. Shettles that if his theory was valid, then artificial insemination should produce significantly more boys than girls. The reason for this is that artificial insemination is usually timed to coincide as closely as possible with ovulation, so as to maximize the chances of conception occurring. (Artificial insemination can be costly and somewhat traumatizing to some women; thus doctors try to achieve pregnancy via artificial insemination with as few attempts as possible.) Prior to any artificial insemination, the conscientious doctor does everything he can to pinpoint as closely as possible the woman's ovulation time.

Dr. Shettles looked at the records of thousands of women who conceived by artificial insemination. He found that about 160 boys were conceived for every 100 girls! Many others have confirmed this fact. Artificial insemination does, indeed, result in a preponderance of male offspring, just as would be predicted by the Shettles theory.

Help at Last

Armed with all this information, Dr. Shettles cautiously began telling patients who asked for help with sex selection that he *might* be able to assist some of them at long last. He also began working, about this time, with the late Dr. Sophia Kleegman, a pioneer in artificial insemination research and a professor of gynecology at New York University's Medical School as well as director of its infertility clinic. Dr. Kleegman sent samples of her patients' sperm to Dr. Shettles for analysis. Both she and Dr. Shettles were soon reporting considerable success in helping couples select the sex of their children. By adding various manipulations to the timing of intercourse, Dr. Shettles found that patients, even without artificial insemination, could conceive children of the sex they wanted at least 75 percent of the time. (The success rate for boys was higher than for girls.)

The specific details of Dr. Shettles' complete methodology, in its updated form, will be explained later in this book. For now, let us take a quick look at the basics, as Dr. Shettles delineated them in a scientific paper published in the September 1970 issue of the *International Journal of Gynaecology and Obstetrics.* Here are some excerpts from that paper:

The difference in shape and size, as well as the correspondence of the overall ratio of sperm type with the conception rate by sex, sug-

gest that other factors are operating, as well as pure numbers. Speed is one such factor and would seem to favor the smaller Y-bearing sperm. Since these are of less mass than the larger X-bearing sperm, they should be able to migrate through the reproductive secretions at the time of ovulation at a greater speed with the same amount of energy, thus making one of them more likely to effect fertilization. When tested in a capillary tube filled with ovulation cervical mucus over a distance of one foot, the small-headed sperm invariably wins the race.

Continence [meaning abstinence from intercourse] or lack thereof is another factor which could favor one or the other type, depending upon circumstances. . . . Continence is associated with an increased frequency of round-heads. Oligospermia [low sperm count] is associated with female offspring. In men with sperm counts of 20 million cc and under [20 million sperm per cubic centimeter of fluid], the likelihood of female offspring varies inversely with the count. With a sperm count of a million or less, only female offspring resulted. . . . This is indicative of the X-chromosome-bearing [female-producing] sperm being the survival of the fittest.

A third factor is longevity, which seems to favor the X-bearing sperm. When the egg is ready for fertilization, this factor may be unimportant, but it is possible for fertilization to occur by a robust sperm which has survived over a period of days within the tube.

Interrelated with the above is different environment within the cervix before and at the time of ovulation. At the time of ovulation the cervical mucus is, among other things, most abundant, most alkaline, of lowest viscosity, and most conducive to sperm penetration and survival. In contrast, the more acid environment within the cervix until a day or so before ovulation is unfavorable to sperm. During this time only the more fit sperm have a chance for survival. The potential to have male and female offspring obviously varies greatly among men. Utilization of each lot of reproductive talents, so to speak, is governed greatly by the timing of coitus in relation to ovulation.

Notice the introduction of a factor not previously discussed: sperm count. Men with low sperm counts, Dr. Shettles has found, tend, if they are not actually infertile because of their low counts, to have more girls than boys. There is nothing mysterious about this. The factors that result in low sperm counts in the first place (various diseases, exposure to certain toxins, excessive heat, stress, etc.) tend to take their toll first on the weaker, boy-producing sperm. The survivors tend largely to be the more robust, female-producing sperm.

Now you know some of the basics of the Shettles method. But how scientific is that method? Have others confirmed it or disputed it? The answers to these questions are in the next chapter.

How Much Scientific Support Is There for the Shettles Method?

We have stated that we believe the Shettles method is the sex-selection technique best supported by available scientific evidence. The method is based, in large part, on these observations:

1. The male-producing, Y-bearing sperm are smaller than the female-producing, X-bearing sperm.
2. The Y sperm can, under ideal conditions, move more quickly than the X sperm.
3. The X sperm are hardier than the Y sperm and are more resistant to various forms of stress.

On the basis of those observations, Dr. Shettles has suggested that:

1. The faster, boy-producing sperm will be most likely to reach the egg first and fertilize it when intercourse occurs at or near the time of ovulation, when the physiological secretions are most alkaline and thus most favorable for sperm penetration.
2. The larger and more resistant, girl-producing sperm will be most likely to fertilize the egg when intercourse takes place more than a day in advance of ovulation, when the secretions are still more acidic and thus most likely to eliminate the less resistant Y sperm.

The Data on Size and Speed

As related in the preceding chapter, Dr. Shettles has tested his hypotheses on the relative speed and size of the two types of sperm. He has done this in a variety of ways. He has observed the behavior of the two types of sperm *in vitro*—that is, in laboratory containers, under varying conditions. He has permitted sperm to "race" through secretions in capillary tubes, and he has employed a special fluorochrome quinacrine dye that causes the tiny Y chromosome to "light up" to help him confirm that the male-producing, Y-bearing sperm are indeed most often the winners of those races in which the secretions are most like those that prevail at or near the time of ovulation. The same techniques have permitted him to confirm that the larger, female-producing sperm survive longer in more acidic secretions, such as those that prevail in the days before ovulation.

More important yet, Dr. Shettles has developed methods for retrieving sperm as they progress through the female reproductive tract. These *in vivo* experiments (those that show what is happening in the living body) are more persuasive than those that are done *in vitro* ("in glass," meaning in lab containers). By retrieving sperm samples at varying times after intercourse, Dr. Shettles can determine whether one type of sperm tends to have more of its kind "out in front" at any given time and under conditions that can be more reliably related to ovulation time.

What Dr. Shettles has found, in the course of these *in vivo* experiments, is that where intercourse was made to coincide with laboratory-confirmed ovulation, the male-producing sperm were indeed the front-runners. The alkaline secretions that prevail at this time were, just as the Shettles theory had predicted, ideal for penetration by the Y sperm. And when intercourse was purposely timed to occur during preovulatory, more acidic phases of the cycle, a preponderance of X sperm were found to have penetrated into those portions of the reproductive tract farthest from the opening of the vagina. Again, Dr. Shettles used the fluorochrome quinacrine dye to help confirm his observations.

Some *have* sought to dispute the claim that Y sperm can move more quickly than X sperm. But, as Dr. Shettles pointed out in a letter published in *Fertility and Sterility,* the contrary data have been based on tests in which sperm have been placed in solutions *other* than those that prevail within the female reproductive tract. Some of these have not even approximated the relative acidity/alkalinity and other biochemical factors that pertain to the *in vivo* reproductive cycle. But even where the pH (relative acidity/alkalinity) has been approximated, Dr. Shettles observed in the publication cited above:

> *It is very important to note that spermatozoal behavior in ovulatory cervical secretion is entirely different from that in the [artificial] chemical solution of the same pH. In the cervical mucus at ovulation time, the amplitude of the spermatozoal movement is decreased and the frequency increased, with the movement being directional with interrupted thrusts forward. . . . In [the lab solution some others have reported using] the movement is more haphazard and random in type.*

In any event, some other researchers, using ovulatory secretions, *have* confirmed Dr. Shettles' observations. Writing in the *American Journal of Obstetrics and Gynecology,* Dr. Shettles cited some of that confirmatory evidence:

> *Rohde and colleagues, in studies of progressive sperm motility in the cervical mucus of women, found a fraction that was rich in Y-bearing sperm in the frontal zone of sperm migration, identified by the quinacrine staining method. Using an* in vitro *penetration test, Kaiser and associates reported similar results. Roberts' observations on gravitational separation of motile sperm substantiated that the Y sperm have greater motility and less mass. Ericsson and co-workers reported isolated fractions rich in Y-bearing human sperm, with the use of progressive sperm motility in bovine serum albumin solutions of various densities, in which the fluorochrome quinacrine stain and fluorescent microscopy were employed.*

Dr. Ericsson (a Ph.D.) has developed a method of concentrating male-producing sperm in the laboratory, based on their faster swimming abilities. Dr. Ericsson, who has patented his techniques so that others must pay to use them, has declared that his sex-selection methods, which work only for boys and require artificial insemination, are in no way derived from the findings of Dr. Shettles. Dr. Shettles merely notes that he published on the faster swimming ability of the Y sperm before Dr. Ericsson did.

Dr. Ericsson has also claimed that Dr. Shettles "lacks data" to support his theory. That is patently untrue, as we are documenting in this chapter. Furthermore, it is a particularly hollow criticism coming from Dr. Ericsson, who has been criticized by medical authorities for "premature publication of findings," without adequate data to support *his* claims.*

As the Ericsson technology has been changing over the years, it is difficult to know precisely what factors the sperm are being subjected to in his separation techniques. Separation techniques that try to use the faster swimming ability of the Y sperm are on the right track, but the solutions and centrifugations to which the sperm are subjected have no parallel in nature—and so what happens to them in this process is open to question.

In 1986 a group of researchers (see the next chapter for more details) attempted to reproduce some of Dr. Ericsson's results, without success. His procedure for "Y-enrichment" (for male sex preselection) actually resulted in samples with greater numbers of female-producing sperm! Hence, Dr. Shettles has concluded that, to the extent that Ericsson has been successful, his results are due more to the timing of intercourse (as close to ovulation as possible for male offspring) than to the artificial separation procedure employed by Dr. Ericsson and his coworkers.

In addition, however, it should be noted that there are serious shortcomings to techniques that require artificial insemination, as we will demonstrate in more detail in the next chapter. At present, the success rate obtainable through these techniques is no better than that reported

---

*See *OBGYN News,* April 1, 1979, "Sex-Determination Study Viewed Skeptically."

by Dr. Shettles and other independent researchers who have *not* resorted to costly, time-consuming, and highly mechanical artificial insemination.

There have been many additional confirmations with respect to the smaller size and faster swimming ability of the boy-producing Y sperm. Even slight differences in the DNA content of X and Y sperm have been shown to result in significantly different motility and velocity. This has been shown to be true both *in vivo* and *in vitro.**

A Japanese researcher reported success in preselecting a small number of female offspring using a separation technique that relied on this slight difference in DNA content. The heavier, female-producing sperm were concentrated in a lower layer of fluid after being spun in a centrifuge. Although this study requires confirmation, the finding, again, is not unexpected—and is in concert with the fundamentals upon which the Shettles method is based.

Dr. B. C. Bhattacharya, the noted Indian zoologist, observed some years ago that farmers in his country preferred bringing their cows in for artificial insemination late in the day, rather than earlier, because, they claimed, sunset inseminations were more likely to result in the bull calves that they particularly desired. Dr. Bhattacharya began studying the situation and found out that the farmers were absolutely right. He ultimately concluded that this was because the two types of sperm settle in their storage containers at different rates. The heavier, female-producing sperm, it seemed, would settle to the bottom first. By the end of the day, the sperm that would tend to be at the top and all that remained for use would be predominantly of the lighter, male-producing variety.

At the famed Max Planck Institute in Germany, Dr. Bhattacharya later experimented with rabbit sperm that was permitted to settle under similar circumstances. When the sperm were cooled and thus prevented from swimming about, separation of the two types readily

---

*See the Bibliography for a paper related to this by A. M. Roberts. Also see one by P. L. Pearson, J. P. M. Geraedts, and I. H. Pawlowitski, related to the larger surface area of X sperm.

occurred. The heavier, female-producing sperm sank to the bottom. Rabbits inseminated with sperm from the bottom layer produced female offspring 72 percent of the time. Rabbits impregnated with sperm from the upper zones had male offspring 78 percent of the time.

Two other researchers, E. Schilling of Max Planck and P. Schmid of Zurich, later confirmed many of Dr. Bhattacharya's findings. And they showed, among other things, that only small differences in the size of sperm can result in differences in velocity of as much as 28 percent in some species. More recently, Dr. Alan Barr of Rensselaer Polytechnic Institute produced a doctoral dissertation called "Spermatozoan Head Shape: A Theoretical Analysis." He, too, found that the head of the X sperm is larger than that of the Y sperm. He noted also that the Y-bearing sperm is faster than the X.

In late 2005, in the December 17 issue of the *British Medical Journal,* a large study again confirmed that the differential in the size of X- and Y-bearing sperm is real and that this difference affects conception outcome. Again, it was noted that the Y-bearing sperm are smaller and faster and better swimmers in the viscous cervical mucus.

"The best-supported hypothesis," said lead researcher Dr. Luc J. M. Smits, affiliated with the University of Maastricht in the Netherlands, "is that the Y chromosome is lighter than the X chromosome." In viscous fluids such as the cervical mucus, he and his colleagues found that the Y chromosomes were fleeter, just as Dr. Shettles had reported on numerous occasions many years before.

Support for the Idea That the X Sperm Is Hardier

The evidence supporting Dr. Shettles' hypothesis that the X sperm is more resistant to stress is less direct but nonetheless highly suggestive. Dr. Shettles has found, in his infertility work, that men with low sperm counts tend to have a preponderance of X-bearing sperm. He theorizes that many of the factors that are known to lower sperm counts, such as heat, certain drugs, toxic substances, even psychological stress, destroy the smaller, more fragile Y-bearing sperm first.

OBGYN News, October 15–31, 1982, reported on a study of licensed undersea divers in Australia which revealed that their offspring are predominantly female (eighty-five daughters compared with forty-five sons). The chief investigator in this study called this difference "very significant" and noted that other studies had previously shown that pilots of high-performance military aircraft also have far more daughters than sons. Both deep-sea divers and high-altitude pilots are, of course, subject to unusual environmental stresses in the form of radically different, and shifting, atmospheric pressures, altered oxygen requirements, possible radiation (in the case of the high fliers), and possible excessive scrotal heat (from tight-fitting wet suits, flight suits, and etc.). The fact that these men father so many daughters meshes with Dr. Shettles' hypothesis on the "greater staying power" of the hardier, X-bearing, female-producing sperm.

In 1988, *Discover* magazine, in an article called "Girls from Space," reported on yet another study showing that high-altitude flying—this time of the sort experienced by astronauts as well as tactical pilots—results in a preponderance of female offspring. The study, conducted by Bert Little, a geneticist at the University of Texas Southwestern Medical Center in Dallas, found that tactical pilots and astronauts exposed to "high G" accelerative forces fathered far more daughters than did pilots who amble through the skies at more leisurely paces. The difference was a highly significant 60 percent to 49 percent female births.

Researcher Little concluded that the higher G forces speed up metabolism in the cells of the body, including the sperm cells, and that the smaller Y sperm burn out before the larger X sperm do. The tendency to produce more female offspring definitely didn't have a genetic basis because, as Little explained, "If a pilot isn't exposed to G forces for about ninety days, then his probability of having a girl drops back down to fifty-fifty." Little noted that high-G pilots and astronauts—and he studied 62 of them in his comparison with 220 other men not exposed to high Gs—also suffer from oxygen deprivation and exposure to ionizing radiation, factors that he believes may also negatively impact the weaker Y sperm.

In 1995, we received a letter from a woman who wondered whether significant exposure to computer terminals might not also predispose men to fathering daughters. Her data is purely anecdotal, but it is certainly worth further investigation. She wrote about a preponderance of female offspring among computer operators she knows. She noted that many of them have discussed this phenomenon among themselves. She ends her letter on a wry note: "Sure wish I could get my husband in front of a computer for just one day. We have three boys!"

Well, if there is anything to this, it would certainly take more than a day. If any of you have further information on this or know men who have worked at computer terminals for prolonged periods, please let us know if you discern any patterns. This may encourage others to do some serious scientific inquiry into the topic.

Other studies have shown that schizophrenics and certain drug abusers also father more females than males. Men who are anesthesiologists have similarly been reported to father primarily girls. After a study in Copenhagen announced this finding some years back, Dr. Shettles spoke with an anesthesiologist colleague who said he could confirm it from his own informal survey of the many anesthesiologists he knew personally. A count of the boys and girls they had collectively fathered revealed a ratio of nearly four girls for every boy! It is known that anesthesiologists are exposed to many toxic substances in the course of their work.

There have been a *few* studies indicating that *certain* diseases seem to result in more male than female offspring among those afflicted. *Usually,* however, the opposite is reported. For example, a group of individuals with non-Hodgkin's lymphoma were studied. Only 36 of the 190 children born to this group were male. That's an astonishingly large difference.* Others have reported that various viral infections can similarly result in more female offspring.

In the February 1978 issue of *Medical Aspects of Human Sexuality,* Dr. Shettles noted that the hardier X sperm are more likely to prevail in a number of other circumstances, as well. Babies conceived as a result of the *failure* of the rhythm birth-control technique are, he observed, predominantly female. So are babies who result from the failure

---

*See *Research in Reproduction,* April 1982.

of birth-control methods that employ acid foams and jellies, used either alone or in conjunction with diaphragms. The reason for this, he is convinced, is that the X-bearing sperm are more likely to survive the acidic environment. The excessive number of girls resulting from rhythm failures can also be attributed to an acid environment. The rhythm failures often occur when women, who may very well have pinpointed their ovulation dates, decide to take a chance and have intercourse two, three, or four days ahead of the expected ovulation time. Most of the sperm *won't* survive until ovulation, but if any *do* survive, they are most likely going to be of the tougher, more resistant, girl variety. That's what the Shettles theory predicts, and that appears to be precisely what happens.

Some have wondered why so many "test-tube babies" have also been girls. These are babies born as a result of eggs being fertilized in laboratory containers, cultured there for a few days, and then implanted into the wombs of women who carry them to term. This procedure, for which Dr. Shettles did much of the pioneering work, has become an increasingly popular means of treating certain forms of female infertility. In it, women who produce viable eggs but have blocked or missing fallopian tubes have their eggs removed. These eggs are fertilized in the lab, using the husbands' sperm, and then reimplanted into the women's wombs at the appropriate time. Since 1978 thousands of babies have been conceived in this fashion.

We do not know the exact number of boys and girls who have been born as a result of these techniques, but various reports have discussed the greater number of girls being conceived at some of the clinics involved in this work. The first nine babies born as a result of this technique in Australia, for example, included *eight girls and one boy.* In a letter to the *American Journal of Obstetrics and Gynecology,* Dr. Shettles speculated that the rigorous preparation of sperm used in the test-tube fertilizations might impose stresses that selectively favor the hardier, X-bearing sperm. The sperm cells that are used in these procedures undergo "washings," suspensions in various fluids, centrifugation, and so on.

Heat is the form of "stress" that, more than anything else, seems to affect sperm counts. And if it is true, as Dr. Shettles, Dr. Kleegman, and others have observed, that high sperm counts favor male conceptions and low sperm counts favor female conceptions, then "keeping it cool," as Dr. Shettles advises his patients, can be important not only for maintaining optimal fertility but also for increasing chances of conceiving sons.

In general, researchers have found that—to quote from a 1990 *New England Journal of Medicine* study—"semen quality deteriorates during the summer. This phenomenon may account at least in part for the reduction in the birth rate during the spring in regions with warm climates." These researchers found that sperm concentrations, total sperm counts, and the concentration of healthy, motile sperm decreased, in their study group of 131 men, by astonishing amounts—by 24 to 32 percent in each category. (Apart from issues related to sex selection, this study tells those couples who have been having trouble conceiving that winter is a better time to try than summer.)

That heat does, in fact, lower sperm counts is no longer in dispute. It hasn't been proved that lower sperm counts result in more female offspring, but the circumstantial evidence strongly suggests that it does. We've already discussed some of that evidence. There's the possibility that even *climate* may have an influence on the sex of offspring. A Canadian geneticist, Dr. Herman M. Slatis of McGill University in Montreal, has claimed that more boys are born in summer than in winter. Others have claimed that men who live in cold climes tend, in general, to father more boys than girls. There is nothing necessarily contradictory about these findings since Dr. Slatis' "boys of summer" would have been conceived in the cool of fall or winter.

Direct and Independent Confirmations

So far we have been talking about evidence that indirectly or partially supports the Shettles method. We will also cite, in that category, the artificial insemination data, since it is reflective of what happens when insemination is made to coincide as closely as possible with ovulation. (As noted earlier, the result is a highly significant preponderance of male births.) But in addition to all of this data, much of which is very important in its own right, is there *direct, independent* confirmation of the Shettles method? In other words, have other researchers tried the techniques, for the express purpose of selecting sex, and obtained similar results?

The answer is *yes.*

One of the earliest direct confirming reports was from Dr. Franciszek Benendo, whose scientific paper in the journal *Polish Endocrinology* was subsequently summarized and reported on in the August 13, 1972, issue of *Medical World News.* The article begins:

> *About a year and a half ago, Dr. Landrum B. Shettles, at Columbia University College of Physicians and Surgeons, proposed a regimen for prospective parents who want to predetermine the sex of their offspring. His prescription was based on such factors as the pH of the cervical and vaginal mucus at various times in the woman's menstrual cycle and controversial theories about differences in the size and shape of X- and Y-chromosome-bearing spermatozoa. He suggested that a major element in determining the conceptus' sex is the time of intercourse with respect to ovulation.*
>
> *Support for part of Dr. Shettles' thesis appears now in a study by a European physician. While he makes no comment on the presumed physiologic basis of the New York obstetrician's regimen, Dr. Franciszek Benendo of the County Hospital in Plonsk, Poland, has confirmed that the timing of coitus that leads to fertilization has a profound influence on the offspring of the sex.*
>
> *Dr. Benendo studied 322 couples in whom the date of fertilizing intercourse and the date of ovulation could be fixed. His first group consisted of 156 married couples "in whom the solitary sexual contact usually took place two to five days before the term of ovulation." In the second group were 18 couples who had intercourse two days prior to the woman's ovulation. And 148 couples who had coitus in the period from one day preceding to two days following ovulation constituted his third group.*

> *One hundred fifty-seven (157) children—including one pair of twins—were born to couples in the first group. Of these, 133 (84.7 percent) were daughters and 24 were sons. Nine children of each sex were born to the second group. And of the 151 children, including three pairs of twins, born to the third group, 131 were sons (86.8 percent).*

And the article concludes:

> *To explain his findings, Dr. Benendo introduces the concept of biological potential. "It may be assumed," he writes, "that the potential of the Y spermatozoa in the first two days' postinsemination is higher than that of the X spermatozoa. After two days, on the average, the X spermatozoa begin to predominate, and their potential becomes greater than that of the Y spermatozoa." According to his statistics, he notes, the magnitude of the difference of the potential is approximately six to one. Finally, "when coitus takes place on the second day before ovulation, as a result of the potential of the spermatozoa and ova at the time of fertilization, 50 percent of the offspring will be male and 50 percent female."*

Dr. Shettles has been in contact with Dr. Benendo since this early publication. The Polish researcher's results have continued much as originally reported. It should be pointed out that Dr. Benendo followed the vital, timing aspect of the Shettles method but did not incorporate some of its other recommendations.

French physician Dr. B. Seguy of Nice reported, in the *Journal de Gynécologie Obstétrique et Biologie de la Reproduction,* that he had achieved a success rate approaching 80 percent using the Shettles method for conceiving males. What makes Dr. Seguy's work particularly interesting is the fact that the one hundred couples he worked with had all experienced infertility problems due to the women's highly irregular cycles. Some did not ovulate at all when first examined. Ovulation was stimulated and the menstrual cycles of the women regularized by the administration of hormones called human gonado-trophins.

Considerable effort was made to ensure that the time of ovulation could be confirmed in each woman. Only then were the couples allowed to have unrestricted intercourse—and this was made to coincide with the actual time of ovulation. *Seventy-seven* of these previous infertile couples gave birth to boys. It seems highly likely that, had these couples enjoyed normal fertility, the number of boys would have exceeded 90 percent.

Dr. Seguy used all of Dr. Shettles' recommendations for male conceptions except for the alkaline douches. With the careful, laboratory-confirmed timing of ovulation, he felt secretions would be adequately alkaline. Nonetheless, he stated that he believed his success rate might have been even greater had he used the douches. (Dr. Shettles has since discontinued use of the douches; see discussion in subsequent chapters.)

Dr. Cedric S. Vear reported, in the *Medical Journal of Australia,* that he had used the Shettles sex-selection method in ten women who had stated preferences for children of one sex. The results? Ten consecutive successes. Dr. Vear did a statistical analysis and concluded that the possibility of this happening entirely by chance was almost nonexistent.

Writing in the *Journal of Biosocial Science,* Nancy E. Williamson, T. H. Lean, and D. Vengadasalam reported on a sex-selection experiment in Singapore. Women were invited to join a sex-selection clinic and use the Shettles method. The experiment was declared unsuccessful, in large part because the women were not highly motivated. This is not surprising, since these women were not actively seeking out a sex-selection method but, rather, were recruited for the purpose of trying one out. Most of the women did not apply themselves conscientiously to the task at hand.

The researchers, in calling the experiment a failure, nonetheless acknowledged that only *six* of the women actually used the Shettles method correctly. Four of those six (66 percent) had children of the sex they desired. If there had been a larger sample of truly motivated women, we believe the results would have been much better. Even here, however, that 66 percent success rate cannot be dismissed out of hand. Some have pointed to this study as evidence that the Shettles method does not work. That is unfair and completely unjustified. The experimental design employed in this study was obviously flawed, but to the extent that it yielded any valid data, that data *supports* the Shettles method.

Parents magazine reported ("Your Child's Sex—Can You Choose?" by Lori Martin) that Dr. Michael O'Leary, an obstetrician and gynecologist affiliated with New York University Medical Center, had used a sex-selection method based on Dr. Shettles' findings. Of the approximately one hundred couples who came to Dr. O'Leary with requests for sex-selection assistance, 75 percent had offspring of the sex they desired, *Parents* reported. "I was one of his successes," the author of the article wrote, "giving birth to a boy some twenty months after my first visit."

The article stated: "At the moment, however, Dr. O'Leary's practice doesn't include sex selection. 'I had 800 women waiting to have it done when I quit,' he says. 'If I hadn't quit, I would have had to abandon the rest of my practice completely. When anybody calls us about it, we tell them to call Dr. Shettles.'"

The most overwhelming confirmation of Dr. Shettles' sex-selection methods have come from Japan. More than one hundred Japanese baby doctors flew en masse to the United States to hear Dr. Shettles explain his techniques. Subsequently he was invited to speak before medical gatherings in Japan and did so on two separate occasions. He also addressed medical students at Gunma University's School of Medicine.

Dr. Shiro Sugiyama, a prominent Tokyo obstetrician and gynecologist and director of the Sugiyama Clinic, reported to Dr. Shettles that the Japanese Sex-Selection Research Group is achieving a success rate of approximately *90 percent* using Dr. Shettles' methods. The Research Group includes 450 specialists in obstetrics and gynecology; in past years these doctors have advised thousands of couples who wanted to select the sex of their children.

Dr. Sugiyama informed Dr. Shettles that there is still significant resistance to sex selection among doctors in Japan. But those who have tried it with their own patients, he adds, quickly become enthusiastic about it. Support for the Shettles theory is thus growing steadily in Japan, Dr. Sugiyama states, adding that an increasing number of conferences and symposiums are being devoted to the subject in that country.

We have to point out that though the Japanese work is based on the findings of Dr. Shettles, there are some variations in their procedures *and* the couples involved get close medical supervision to ensure that instructions are carried out accurately. This direct medical help no doubt explains the higher success rate of the Japanese. The Japanese study is ongoing, and the results, Dr. Sugiyama indicates, will eventually be published.

In the next chapter, we examine some other methods of sex selection that have been proposed in recent years. Some of these, though they appear sometimes to be in conflict with the Shettles method, actually confirm various aspects of Dr. Shettles' work.

Other Sex-Selection Methods: How Do They Compare?

Since we last revised this book in 1996, some new developments have emerged in sex-selection technology. They are some new methods, and some old methods have been "repackaged" in an effort to make them look new. Some of the other methods are consistent with the findings of Dr. Shettles, and some conflict. Others, especially some of the new so-called high-tech methods, can be effective, but, as you will see, they are generally no more effective than the Shettles method, they are invasive, they carry significant health risks, and they are very costly. Some are ethically challenged.

First we examine some of the low-tech methods, and then we move to more recent developments on the high-tech front.

The Whelan Method

(Revived by the Young Method)

The Whelan method, which generated brief publicity, is rarely used today. It was based on a hypothesis that became the subject matter of a book by a woman named Elizabeth Whelan. Most of its tenets were in direct conflict with the findings of Dr. Shettles. Whelan claimed a success rate of 68 percent for girls and a 57 percent success rate for girls (barely above what Mother Nature provides) for those who used her method. We maintained from the outset that the assumptions underlying the Whelan method were unsound and often contradictory.

Whelan based her method on the work of a physician in South America (who did not collaborate with her on the book). Dr. Rodrigo Guerrero of the Universidad del Valle in Cali, Colombia, had found, just as Dr. Shettles had, that a decisive majority of children conceived through artificial insemination, where care is taken to inseminate at the time of ovulation, are boys. Dr. Guerrero's artificial insemination data appear to be sound, since they were based on statistics from respected clinics in the United States where record keeping was meticulous and where time of ovulation could be determined reliably.

So far so good. Unfortunately, Dr. Guerrero, using data that we believe were completely *unreliable*, went on to report that in *natural* insemination, via sexual intercourse, fewer boys than girls were conceived at or near the time of ovulation. No convincing explanation was given to explain why there should be this striking and remarkable difference between artificial insemination and natural insemination, a difference that flies in the face of commonsense expectations. Nonetheless, one would still have to take this report seriously—seriously enough, in any event, to look closely at the data on which it is based.

Here's where the trouble begins. Whelan, in our view, did not look with discrimination at that data and went on to extrapolate her method from it, telling her readers to try for boys on the sixth, fifth, and fourth days prior to the day of presumed ovulation. For girls she suggested trying closer to ovulation, either on the day of ovulation or two or three days prior to it. In making these recommendations, Whelan tended to ignore that part of Dr. Guerrero's data that was most sound—the artificial insemination statistics that indicate that intercourse at or near ovulation will produce boys. Instead, she chose to rely on Dr. Guerrero's natural insemination data, data that, unfortunately, was retrospective (the type of data that requires one to "look back" and try to figure out what happened at some point in the past). With retrospective studies, there is no scientific observer on the scene observing events *as they unfold,* making sure that reported events really do occur and that they occur at the times and in the ways reported.

Dr. Guerrero's retrospective data on natural insemination was shaky, at best, for he had to rely, after the fact, on charts kept by women who were, largely, trying to avoid pregnancy through the rhythm birth-control method. He looked at these charts and tried to figure out from them the interval between the last intercourse prior to conception and the probable time of conception itself. This could not be done with any significant reliability even if these were the charts of women with generally regular menstrual cycles. One could not be certain that one had pinpointed the time of ovulation merely by looking at isolated charts. Here the situation was even worse, however, for it must be remembered that many of the women whose charts Dr. Guerrero was trying to divine some months after they had been completed were *failures* in the rhythm method of birth control. The main reason for rhythm failures—and they are frequent—is the woman's inability to correctly monitor her menstrual cycle, an inability that is reflected, inevitably, in *inaccurate charts.*

We see no reason whatever to believe that Dr. Guerrero's natural insemination data have any validity. Others have commented on the weaknesses of Dr. Guerrero's research design. Dr. Robert Glass, of the department of obstetrics and gynecology at the University of California School of Medicine, for example, has noted that the way Dr. Guerrero timed ovulation was inconsistent. Thus, he concludes, it is difficult to tell, with any confidence, when the temperature shift that marks ovulation occurred in these experiments.

Dr. Guerrero, in addition, based his findings in part on women whom he believed had conceived as a result of intercourse that took place "six or more days before ovulation." We doubt that *anybody* conceived as a result of intercourse that far in advance of ovulation (sperm simply can't live that long), but even if a few did, the statistical sample would necessarily be so small that no valid conclusions could be based on it. Indeed, even the journal that published Dr. Guerrero's findings (*New England Journal of Medicine,* November 14, 1974) noted, in an accompanying editorial, that even if the results proved reliable, they would be of little use to anyone, since it would take, the editors concluded, *at least* five years for anyone to conceive a child, of either sex, if intercourse was halted that far in advance of ovulation. We think the editorial writer was being too optimistic; only rarely will a sperm cell survive longer than four days and still be capable of fertilizing an egg. And if one *does* survive that long, it will most likely be of the *girl*-producing variety, not the boy-producing type, because only the larger, girl-producing sperm, with its greater stamina, would likely survive for that period of time.

Dr. Shettles has concluded that those who follow the Whelan method for begetting boys usually will simply fail to conceive a child of either sex. Whelan herself, in one of her more striking contradictions, acknowledges that sperm have a *maximum* survival time of ninety-six hours. That is *four* days. How, then, are couples to become pregnant trying to conceive five and six days before ovulation? Even at four days their chances are going to be slim. As for those who follow Whelan's suggestions for begetting girls, Dr. Shettles notes that Whelan manages to be *both* in concert and in conflict with his recommendations. Those who try for the girl two or three days before ovulation will have in-creased chances of getting one. There he agrees with Whelan. But those who try on the day of ovulation or the day prior to it will most likely have boys. There he differs with Whelan.

Whelan has suggested that the Shettles method is based entirely on the artificial insemination data. That is untrue, as any who have read this far know. But, in fact, the artificial insemination data is very im-portant and certainly far sounder than the natural insemination data of Guerrero. Whelan has been at pains to try to establish that there are sig-nificant differences between natural insemination and artificial insemi-nation and that these differences account for the different results reported by Dr. Guerrero. She argues that sperm that enter the repro-ductive tract in the course of natural insemination have "starting points" that are different from those that enter via artificial insemina-tion. Sperm from natural inseminations, she says, "have to work their way up the vagina." Sperm introduced artificially, on the other hand, are "placed close to the cervix."

That difference is real, to be sure, but there is no reason to believe that it accounts for the discrepancy between the artificial insemination data and the Guerrero data. A badly flawed experimental design, as explained above, is the most likely explanation for that discrepancy. And as for the different "starting points" of the sperm, that is a factor that Dr. Shettles has long taken into account. In his method for begetting sons, for example, Dr. Shettles has always recommended deep penetration by the male at the time of seminal ejaculation, so that sperm will be deposited close to the cervix. For female conceptions, Dr. Shettles has always recommended shallow penetration, so that the sperm have to make their way up through the more acidic environment of the vagina, an environment that the female-producing sperm are better able to tolerate.

We think it very unfortunate that Whelan would make recommendations to the public based on "evidence" of so insufficient a nature. On the other hand, we have been gratified by the many letters we have received from individuals who have clearly perceived the deficiencies and contradictions of the Whelan method. One woman did a particularly impressive job of critiquing the Whelan book in a letter to Whelan, with a copy to Dr. Shettles. With her permission, we include excerpts from that letter:

Dear Ms. Whelan:

Six months ago I purchased Dr. Landrum Shettles' *How to Choose the Sex of Your Baby.* Last week a friend gave me a copy of your *Boy or Girl?* which I have now read. There are many points on which I disagree . . . On page 87 [of the paperback edition of Whelan's book] you state, "sperm may retain their capacity to fertilize an egg for ninety-six hours—or in some rare cases possibly even longer." On page 117 you state, "If you want a boy, have sexual relations on the sixth, fifth and fourth days before the expected day of the temperature rise." On the chart on the bottom of page 109 you even indicate *nine* days before the temperature rise. If the sperm only live a maximum of *four* days, how can the average woman become pregnant *nine* days before the temperature rise?

On page 113 you state, "Although there is not a full consensus on this point, it is generally believed that ovulation takes place two days before the sustained rise in temperature." Our family doctor, my gynecologist, my temperature chart instructions, the instructions with my basal thermometer and all my medical literature refute this statement. They disagree as to the *exact* moment but *do* agree that ovulation occurs either at the bottom of the temperature drop or somewhere between the lowest temperature of the drop and the top of the rise. . . .

Both my boys were conceived on the fourteenth day of my cycle. . . . A friend conceived two boys on the fourteenth day of her 28-day cycle. By following Dr. Shettles' advice on timing for girls, she achieved her girl. That makes five pregnancies I have cited. All four boys were conceived at mid-cycle, at or near ovulation, the time you say should produce girls. The girl occurred after using Dr. Shettles' timing method, the time you say should produce boys. That makes 100 percent success for Dr. Shettles' method and zero success for your method. . . .

On page 38 of his book, Dr. Shettles says, "Intercourse two or three days before the time of ovulation . . . would be likely to yield female offspring." On page 129 of your book, you say, ". . . you might consider that by following the Shettles-Rorvik advice in hopes of having a baby girl you would drastically reduce your chances, because you'd be concentrating coitus three or more days before ovulation, on what we have found are the 'boy days.' " . . . on page 117, you say, "If you want a girl, do just the opposite: avoid sex until two or three days before the expected rise in temperature." This seems to *support* Dr. Shettles' statement. Another contradiction?

. . . The real point of contention between you and Dr. Shettles on the girl timing seems to be *when* ovulation really takes place because both of you say intercourse two to three days before the temperature rise gives the greatest chances for a girl. In your view intercourse is then taking place *at* ovulation and in *his* view intercourse is taking place *before* ovulation. And yet your chart on the bottom of page 109 shows that the greater percentage of girls occurs *at* the temperature shift which refutes the page 117 statement that girls should result from intercourse *two or three days before* the temperature rise! Another contradiction?

If this all seems confusing and contradictory to you, perhaps you can understand my feelings toward your book when I finished reading it; *it* seems confusing and contradictory!

This letter was written by Kathy Brown of Ada, Ohio.

Whelan, incidentally, does follow one aspect of the Shettles method in her sex-selection book. She concludes that, with respect to the alkaline and acidic douches, "There is enough presumptive evidence to advise, 'give it a try.'" In that book, published several years ago, she recommended the same alkaline (baking soda) and acidic (vinegar) douches that Dr. Shettles once did, adding that both are harmless. As Dr. Shettles once did, she suggests the alkaline douche for boy attempts and the acidic douche for girl attempts. She has even recommended, as Dr. Shettles did long before her, that, when trying for the boy, the woman should try to have her orgasm *before* her husband does, in order to increase the alkalinity of the natural secretions and help carry the sperm quickly up into the cervix. And for the girl she recommends, again as Dr. Shettles did years earlier, that the woman avoid orgasm.

We must point out, however, that the douches are of minor importance when compared with the timing factor. And if you follow the Whelan method, we believe that no amount of douching will compensate for the wrong timing she recommends in most instances.

In 1995, just as Whelan seemed to be fading from the scene, J. Martin Young came along to revive her contrary method. Interestingly, Young never acknowledges Whelan (afraid of the competition?) and is either ignorant of our post-1970 work or pretends to be (afraid people will buy our book instead of his?). We find it peculiar that someone who brags about using computer databases to discover all he can about sex preselection would not discover that our 1970 book has undergone substantial revision and updating *five times* since that date!

But, then, we're not impressed with the rest of Young's "research," either. Nor are we impressed with the way Young has "packaged" his material. In what appear to be two self-published paperback books (from a company called Young Ideas), Young outlines his method in two volumes: *How to Have a Boy* and *How to Have a Girl*. We must point out that these two books (which cost $19.95 *each!*) are virtually the same book, word for word. You can boil the differences down on a few pages, and all of Young's material could easily have been contained in one thin volume.

It was suggested to us some years ago that we similarly separate our boy and girl methods into two separate volumes and thus greatly increase our profits. Both we and our new publisher rejected that suggestion on the grounds that it was unnecessary and would be a blatant rip-off of our readers. Moreover, we have long been convinced that it is very important for our readers, no matter whether they are trying for a boy *or* a girl, to be able to study *both* methods *side by side* so that they can reinforce not only what they *should* do but also what they *should not* do. And, of course, many of the readers we hear from plan to try, eventually, for children of *both* sexes—and we do not think it fair to ask them to buy two separate books to get to that goal.

Now on to Young's research. He claims that if you follow his method, based on research conducted not by him but by others, you will have a 51 to 68 percent chance of having a boy or "greater than 60 percent" chance of conceiving a girl. If those numbers don't have you shouting "hurrah," prepare for even worse news. Young not only has no significant clinical experience with sex selection but has based his entire methodology on six studies, conducted between 1974 and 1985, two of which he admits have no statistical significance whatsoever and one of which, as he calculates it, has marginal statistical significance. The remaining three studies—no surprise here—include those of Guerrero, which we have already thoroughly discounted, as have others.

John France and associates published a study in 1984 that Young states contradicts the Shettles method on the issue of timing of intercourse. "Thus," states Young, "following the Shettles' theory actually *decreased* their probability of success." There are a number of problems with this conclusion. France and his group did not follow the Shettles method in several particulars, and so it could not be called a fair test of the method by any stretch of the imagination. And France used retrospective methods to try to determine the day of ovulation. Finally, so few couples in the study actually became pregnant that it would be foolhardy to try to draw any conclusions at all from it.

Young admits that the World Health Organization data he tries to extrapolate from are statistically nonsignificant. He makes the same admission with respect to the work of Barbara Simcock, in which he unsuccessfully seeks to find a refutation of the Shettles method. The trouble is, for those who do not watch carefully what Young writes, these admissions may get lost in the rest of his rhetoric.

And so we come to the sixth study upon which he claims to find support for his method. This is the 1985 study of A. Perez and associates in Santiago, Chile, another of those nonclinical, retrospective "paper" studies in which researchers try to figure out, from the charts of women they often never even have met, on what day they ovulated. These were the charts of women who were using natural family planning. We have already discussed in detail the shortcomings of such studies.

It is no wonder that P. W. Zarutskie and associates, reporting in 1989 on their analysis of these studies in the journal *Fertility and Sterility,* concluded that *all six studies upon which Young relies contain no information that can be useful for sex selection by average couples.*

In conclusion, Young's method has nothing more to commend it than did the Whelan method. Indeed, in at least a few respects, the Whelan method was a little more sensible. Virginia Gilbert, writing at the online site Preconception.com in 2004, interviewed a woman who had investigated both the Shettles method and the Whelan/Young method and concluded that a "30-year-old homemaker in Illinois could be the Shettles Method Poster Mom." She explains how Bethany and her husband followed the techniques outlined in *How to Choose the Sex of Your Baby* when they conceived each of their five children. They got pregnant on the first try four out of five times. One baby took two months to conceive. "And most impressive, this couple achieved the desired gender with all of their kids: Lindsay, Haley, Makenna, Camden (the only boy) and Amrin. . . . When asked how she feels about having employed Shettles' somewhat rigorous techniques, Bethany gives her conception experience a big thumbs-up."

Other Methods Based on Timing and Frequency of Intercourse

After Dr. Guerrero's data was published, a nonmedical British researcher named William James came up with a curious sex-selection hypothesis. James accepted on faith the Guerrero thesis that boys are conceived when intercourse occurs well before ovulation and that girls are conceived when intercourse takes place nearer the time of ovulation. On the basis of this erroneous assumption, James went on to speculate that a woman who has sex every day or nearly every day will most likely have boys. He made an effort to back this up with statistics that claim to show that among first births, there is an excess of boys born to mothers under age twenty-five, while women over thirty-five give birth to more girls. He tried to tie all this together by citing "unpublished data" that he said "powerfully suggest that coital rates [frequency of intercourse] in the first month of marriage are higher than during any subsequent month."

Before you consider basing your sex-selection method on this statistical house of cards, be aware that other researchers have failed to discern the same patterns.* We don't doubt that newlyweds make love more often than do most longer-married couples, at least on average, but we see no real evidence that this results in more male offspring. If, in fact, more male offspring are being born to younger couples, this is likely due to factors other than frequency of intercourse. Younger husbands tend to have higher sperm counts than older husbands; higher sperm counts, as noted earlier in this book, tend to favor male conceptions. Similarly, younger women tend to have more copious and alka-

---

*Note, for example, Dwight T. Janerich's report in the medical journal *Lancet*, April 24, 1971.

line cervical secretions than do older women; those alkaline secretions tend to favor male-producing sperm.

One thing we're convinced of: having intercourse *every* day will result in more girls, not more boys. The reason for this is that when ovulation occurs, a certain number of sperm, ejaculated two or three days earlier, are likely to be lurking high up in the tubes, waiting for the egg. Those hardy survivors are most likely to be girl-producing sperm.

Some years ago an Austrian researcher, Dr. August J. von Boronsini, reported that men who had intercourse far more often than average (he studied some men who still have harems) father far more daughters than sons. This makes sense to us. The sperm counts of such men are likely to be severely taxed by conjugal duty.

Perhaps the most surprising claim James made was that Orthodox Jews living in Israel have more female offspring than non-Jews. He attributed this to the Orthodox Jewish practice of *niddah,* which calls for abstinence from intercourse until a week after the menstrual period, as previously explained. We have speculated that this practice places the first permissible intercourse in each cycle at or near the time of ovulation in most women. Dr. Shettles has argued that this should result in more male offspring, and, in fact, this is what has been reported by various authorities over the years.

Another nonmedical researcher, Susan Harlap,* also studied Orthodox Jews who observe *niddah.* She concluded that more male babies than female were conceived when the first intercourse occurred two days *after* ovulation. This was, however, another retrospective study and one in which the potential for error was enormous.

An editorial comment, by Dr. Joe Leigh Simmons, accompanies the Harlap report, stating:

> *The time of insemination and its relation to ovulation were estimated solely from information obtained after delivery, i.e., nine months after conception. Unfortunately, many pregnant women have difficulty recalling even the approximate dates of their last*

*See Bibliography.

menstrual period, much less other details. [These women did not even have charts to remind them of what may have happened.] Thus, one could question the reliability of the methods used to identify pregnancies resulting from insemination on specific days. Moreover, since most studies indicate that the fertilizable period after ovulation is only 12 to 18 hours, fertilization two days after ovulation seems relatively unlikely.

Dr. Simmons observed that some of the statistics could not be dismissed out of hand but concluded that "the relatively small number of births in this subgroup does not, of course, justify far-reaching conclusions."

Harlap herself acknowledged that when women reported irregular cycles, such as "26 to 30 days," she simply based her estimations on the midpoint of the range given. This, clearly, will result in gross inaccuracies, as will the inevitably faulty recall of many of the women.

It is probably not possible to get an entirely accurate picture of what occurs with respect to Orthodox Jews who observe *niddah*. The validity of the Shettles method is neither sustained nor undermined on the basis of findings related to *niddah*. There are too many unknowns and variables at work here to be able to say anything absolutely conclusive. Nonetheless, it does *appear* to us that those Jews who observe this practice are more likely than not to engage in intercourse for the first time in any given cycle at or near the time of ovulation, which should result in more male than female offspring, just as has been reported by many researchers over the years.

Dr. Jacob Levy of Jerusalem published a paper in the Hebrew-language *Koroth* in 1973 asserting that among Jews who are known to *strictly* observe *niddah*, male offspring do indeed outnumber female offspring. In this population, Dr. Levy reported, there are 130 or more males born for every 100 females. The average birth ratio for non-Orthodox Jews, he stated, is 100 females for every 105 males, the same ratio that prevails in most parts of the world.

Dr. Levy pointed specifically to "the new theory of Shettles" as a logical explanation for this preponderance of males.

One last point before we conclude our discussion of those who have posited sex-selection methods based on timing and frequency of intercourse. William James, discussed above, produced data related to fraternal twins that appear valid. He reported that fraternal twins (the result of two different eggs, released at about the same time, being fertilized by two separate sperm) are almost always of the *same sex*. This provides further evidence that timing of intercourse is of vital importance. If timing weren't important, then one would expect fraternal twins to be of different sexes most of the time, reflecting the roughly fifty-fifty division of the sexes. But, as it happens, it is clear that conditions within the female reproductive tract—and to some extent within the male reproductive system—at any given time largely determine which sex shall be conceived.

In 1995 a group of researchers stated they had found no evidence that timing of intercourse influences the sex of the baby. Their study, however, based on data from women they recruited in the early 1980s, focused primarily on trying to discover at which point in a woman's cycle sexual intercourse was most likely to result in conception, irrespective of the sex of resulting offspring. Not surprisingly, they found that the chances increase the closer intercourse occurs to ovulation. A finding in this study reassuring to us in terms of our girl method was that the age of the sperm does not seem to be an issue in terms of the viability of a resulting pregnancy.

As for the negative finding between timing and sex of offspring, we find the data relied on was insufficient to reach such a conclusion. The researchers had to rely on reports from couples themselves as to timing of intercourse, number of incidents of intercourse, timing of urine samples (from which the researchers tried to deduce time of ovulation)—all subject to erroneous and incomplete reporting, which is very common in such "self-monitored" studies. In addition, only a small subset of the study group could be considered relevant to any findings related to sex of offspring, since most of the women studied were having intercourse every day or nearly every day during the fertile period. These were, after all, women who wanted to get pregnant, not women who had any interest in selecting the sex of their offspring. And so the researchers had to try to sort out those women who had sex only on *one* day of the fertile period in order to try to come to any conclusions about timing of intercourse and sex of offspring. This was a very weak experimental design by any reasonable standard—so weak that no conclusion can be drawn from it, in our view. The preponderance of evidence, already discussed, powerfully supports the timing element in sex preselection.

A couple of other methods bear mentioning, although they not only verge on the absurd but surpass the absurd. One is the Eugen Jonas method, which arose in the 1960s and seems to be enjoying a bit of a revival via the Internet these days. This method is based on what is grandly called "cosmobiology"—in short, on phases of the sun and the moon. The position of the moon, it is claimed, is particularly powerful in influencing gender outcome. Thus purveyors of this method sell couples charts for $50 to $400. In *The Skeptic's Dictionary* (2005), Robert Todd Carroll observes that "one reason for doubting any of these charts will work is that there have been several studies that looked for a significant correlation between phases of the moon and birth rate. All have come up barren." He cites extensive studies in Italy, France, and elsewhere that showed no correlation between lunar and solar activities and fertility or gender outcomes. Carroll concludes: "For a medical doctor to claim that mental retardation and other birth defects are due to heavenly bodies is a sign of lunacy. What Jonas is selling is hope packaged in scientifically sounding mumbo-jumbo. He may well believe what he is doing, but in my view he should be arrested for fraud."

Another method being touted online bears looking out for—that is, *avoiding*. It is offered by something called "The Birth Planning Center" and its "Baby Gender Selection Program" (although these names can change overnight for marketing purposes). The site presents a lot of hype: "Hey future parents! Finally discover how to successfully choose the sex of your baby with an impressive and proven 94 per cent success rate!" Then it trots out testimonials but no scientific data about the actual method. Order forms keep cropping up, of course, and the reader is reassured that "my technique will skyrocket your chance of conceiving the boy or girl you want to the roof." And why should you believe all this? Because, you are told, "We have a trustable reputation."

Our method has been ripped off without credit many times, so we were relieved to see that in the promotional material for this online offering, the purveyor proclaims that the "incredible Technique . . . has nothing to do with Dr. Shettles' method." The only details revealed about the method make the completely unfounded claim that "for more than nine years now, it has been proven over and over again . . . that each and every month there are specific and limited DATES on which you really can successfully conceive the gender of your choice. . . . it has nothing to do with the Timing between ovulation and intercourse Theory. You wouldn't believe the insane amount of couples who tried that and failed."

We would be laughing out loud except that we know that some people actually get taken in by this kind of garbage. Remember what our mothers taught us: When surfing the Web, beware of sharks.

Finally, and with some irritation, we must mention a couple of books that have come to our attention in 2006. One of them is called *Choose the Sex of Your Baby: The Natural Way,* by Hazel Chesterman-Phillips, published in England. This author describes a method that she says just happens, happily for her sake, to coincide with the Shettles method. Hmmm. We are investigating. Meanwhile, we suggest readers in the United Kingdom go to the original source of the Shettles method—the book you are reading right now. When others "filter" our method, apart from possible legal issues, details often tend to get mangled.

The other book is called *Baby Girl or Baby Boy,* by Mark Moore, Lisa Moore, and Jeff Parker. This book, published in 2004, also uses the primary elements of the Shettles method. We plan to investigate this one too, as to the status of our copyright.

Sperm-Separation Techniques and Artificial Insemination

Sperm-separation techniques are being developed, based primarily on the different swimming speeds of the two types of sperm. Various researchers have confirmed Dr. Shettles' observation that the smaller, boy-producing sperm swim more rapidly than the larger, female-producing sperm. In these techniques, sperm are placed in columns of increasingly concentrated albumin and other fluids and permitted to swim through them. The male-producing sperm swim through first and can thus be concentrated and used in artificial insemination procedures.

Unfortunately, these techniques, used by Dr. Ericsson (mentioned earlier) and by Dr. Paul W. Dmowski of Michael Reese Hospital and Medical Center in Chicago and others, are not perfect. Some female-producing sperm are still present in the concentrate; so while the chances for male conceptions are increased, they are not guaranteed. The sperm that get left behind in these separation procedures, moreover, are a mixture of both male- and female-producing sperm; thus the technique can be used only to increase chances of getting sons. Other drawbacks include the cost and inconvenience of the separation procedures and of artificial insemination. Artificial insemination does not always succeed on the first attempt; often it has to be repeated two, three, or even more times before conception occurs. This, of course, increases the costs significantly.

Some practitioners of sperm separation speak of an 80 percent success rate. This is misleading for a number of reasons. To illustrate, let's look at a sample of ninety-nine women who want sons by this method. It is likely, based upon actual cases reported in the literature, that about *a third* of these women will drop out of the program after an average of three unsuccessful efforts to conceive a child of either sex via artificial insemination. Of the remaining sixty-six women, perhaps as many as 70 percent will sooner or later conceive. Of those forty-six women, let us accept the claim that 80 percent will have sons. Eighty percent of forty-six is thirty-seven. But, remember, we *started* with ninety-nine women; only thirty-seven of those got what they wanted. That, in reality, is not an 80 percent success rate but, rather, a *37 percent* success rate!

Dr. Shettles himself has experimented with artificial insemination as a means of selecting sex. He hears from a small number of couples who *think* they would like to use it to beget sons. They think so until they try it. Then many complain of the "mechanical" nature of it, the "artificiality," the "lack of spontaneity," the "unaesthetic" nature of it, and so on. *Infertile* couples are more likely to adapt to it because they see it, in some circumstances, as their *only* chance of conceiving.

Yet, in 1986, the American Fertility Society noted that artificial insemination "may make reproduction excessively technologic by separating procreation from sexual expression."

To a certain extent, *all* forms of sex selection meet with *some* resistance because they all entail some inconvenience and artificiality. But the sort of artificiality entailed by the Shettles method is minor compared with that of artificial insemination. Demographer Anne Pebley of Princeton University's Office of Population Research, which has conducted extensive studies into the sociological aspects of sex selection, made this observation about couples who view sex-selection techniques somewhat negatively: "If they were told that artificial insemination were necessary, they may be even less likely to approve."

As noted in the previous chapter, some research has raised new questions about the mode by which sperm-separation techniques work. Researchers at the Lawrence Livermore National Laboratory of the University of California and the East Bay Fertility Obstetrics/Gynecology Medical Group in Berkeley, California, for example, tried to produce Y-enriched (male-producing) sperm using the Ericsson procedure and failed.

This was a carefully constructed and controlled set of experiments in which, according to the researchers' report (published in the October 1986 issue of *Fertility and Sterility*), "all [semen] samples were processed by an individual very experienced with the Ericsson serum albumin isolation procedure." One sample, in fact, was processed by Dr. Ericsson himself. And the East Bay Fertility Obstetrics/Gynecology laboratory is one of the centers licensed to use the Ericsson procedure to produce male offspring (with a success rate, as of October 1986, of 72 percent: forty-nine males and twenty females out of sixty-nine term pregnancies resulting from these efforts). There has been no increase in this success rate.

So here you have a situation where even those who have reported some success with the Ericsson method are at a loss to explain *why* they have succeeded. These researchers found that the separation procedures that were supposed to result in Y-enriched sperm actually resulted in samples with slightly greater numbers of female-producing, X sperm! They hypothesize various possible reasons for this remarkable situation, none of which seems to convince even them and none of which convinces us.

They do acknowledge that there are "major differences" between this artificial approach to conception and the natural one—but they dance around the most obvious explanation for a finding that could only have sorely embarrassed those using the Ericsson method. That explanation is *timing,* the factor that Dr. Shettles has always held to be most important in sex preselection.

This report suggests that, after all of their elaborate and costly manipulation of sperm, those using the Ericsson method are introducing a mixture of X and Y sperm not much different (in relative numbers) from those that are introduced in the normal course of events. Great care is taken, however, to introduce them as close to the time of ovulation as possible to maximize chances for conception. If this carefully designed study was accurate—and there are a great many reasons to believe that it was—then it is fair to suggest that the Ericsson method is no better than ordinary artificial insemination—and that it works to the extent that it does because of the *timing of insemination.*

Dr. Shettles made this point when he wrote, in the May 1987 issue of *OB/GYN World,* that "apparently, it is the timing of the artificial insemination that is the critical factor. With Ericsson's protocol, the processed sperm is artificially inseminated into the ovulatory cervical mucus; the secretion at this time affords the optimal milieu for the X and Y sperm to manifest their normal differential in migratory rates, as reported by Rohde et al., Shettles and Kaiser et al. These workers observed that the Y sperm swim the fastest in the ovulation phase cervical mucus, i.e., without any processing."

The Japanese have come up with yet another variation on sperm-separation techniques, this one aimed at producing X-enriched (female-producing) sperm. "New Way Devised to Pick Child's Sex," declared the headline in the September 23, 1987, issue of *The New York Times*. This "new" method, the article went on to state, "is based on differences between sperm carrying X chromosomes, which produce girls, and Y chromosomes [which produce boys]. The DNA content of X-bearing sperm has been found 2.7 percent greater than that of Y-bearing sperm."

Sound familiar?

What may be new about the Japanese work (under the direction of Dr. Rihachi Iisuka at Keio University in Tokyo) is the sperm-separation technology. It's possible that the Japanese really have found a way to concentrate X-bearing sperm, using differences Dr. Shettles discovered many years ago. They are claiming a high success rate but, again, bear in mind that this procedure requires artificial insemination, which itself has a substantial failure rate. Bear in mind also that no one has yet confirmed the Japanese work and that it is based on an extremely small sample—only six cases. You should be aware, as well, that the Keio researchers are applying the procedure *only* to couples who cannot risk having male children because of sex-linked hereditary diseases that afflict their male offspring, such as hemophilia.

In 1987, we received a letter from a couple seeking to have a daughter. They had been advised that if they had a male child, he might have hemophilia. They went to a clinic claiming to have a method of producing "female-enriched" sperm—that is, semen containing mostly X-producing sperm. The woman underwent artificial insemination with sperm processed by this clinic but did not become pregnant.

A follow-up attempt was not made. The clinic closed because, as the woman explained it, they "were getting male babies even when trying for girls." The writer added: "I can understand this result [the male babies] after reading your book. Insemination was done at the time of ovulation using an ovulation-predictor kit." Again, *timing* was overriding everything else.

The doctor who had been treating this couple, the woman wrote, "suggested we might try on our own to produce a girl by not having intercourse any closer to ovulation than two days. His suggestion seems to be consistent with your procedure." At the time she was writing she had not yet become pregnant.

She concluded her letter: "My fear, however, is that this procedure is acceptable for a couple who simply would *like* a child of a certain sex, but that it may not be acceptable for a couple who so desperately *need* a female for the child's health."

We totally agree. We never recommend *any* method of sex selection to those with sex-linked hereditary diseases. Such couples must decide for themselves, in consultation with their own doctors, whether to try to conceive.

The latest word on sperm-separation techniques continues to be more negative than positive. A group of researchers at the University of Adelaide and the Queen Elizabeth Hospital in Australia reported that the separation techniques they examined did not result in reliable enrichment of either X- or Y-bearing sperm.

The most deplorable aspect of some of the claims that have attended the sperm-separation sex-preselection procedures is the representation, whether explicit or implicit, that there at last is a *surefire,* fail-safe method of sex selection. There is no such thing—and we do not anticipate that there will be one for many years. (See the Afterword, "Sex Selection in the Near and Distant Future.")

The bottom line, however, is that most couples can achieve results as good as, or better than, those being reported for artificial insemination *without* using it. We'll have a bit more to say on this subject later in this book.

Emerging High-Tech Sex-Selection Methods

In addition to the older spinning or gradient method employed by Ericsson and others, some newer sex-selection methods are emerging. One of these is called flow cytometry (brand name MicroSort). In this method, special fluorescent dye is used to identify X-bearing sperm. Because, as we have explained, the girl-producing sperm are larger and contain more DNA, they absorb more dye and thus can be sorted out microscopically. Sperm of the desired gender can then be introduced via artificial insemination.

The practitioners of flow cytometry claim a success rate of 70 to 90 percent, but, again, a number of caveats are in order. The statistics get distorted, just as they do with the spinning or gradient methodology, since a significant number of women fail to become pregnant at all via this method, so the bottom-line success rate is actually quite a bit lower. This method typically costs in the neighborhood of $3,000 per treatment cycle. In studies related to this method, the pregnancy rate for each artificial insemination cycle is only 16.6 percent. If fertility drugs are added, you will pay an additional $2,000 or more. Also, the method is still considered experimental and has been tried in only a few hundred women. It has not received FDA approval. Candidates must be married, aged thirty-nine or younger, and have at least one child of the gender opposite that desired.

Another newer method receiving much more press attention of late is preimplantation genetic diagnosis (PGD). If you listen to the hype, PGD is the ultimate "breakthrough" in gender selection: safe, easy, and virtually 100 percent successful. In reality, it is a highly complex process and carries with it a number of potential perils as well as very high cost. And it is not nearly as successful as the hype would lead you to believe.

PGD grows out of research that was originally intended to detect genetic defects and anomalies in the embryo, thus offering the option of terminating the pregnancy when defects were judged serious enough to warrant this. But since the genetic analysis also revealed the gender of the embryo, demand began to grow to use the technique for sex selection. Most medical authorities, medical organizations, and bioethicists continue to oppose the use of PGD for this purpose because it entails discarding embryos of "unwanted" gender. Nonetheless, a few enterprises are offering the method to couples for sex selection.

Those contemplating it need to realize at the outset that PGD requires *in vitro* fertilization (IVF). An anesthetized woman's eggs are extracted via a syringe inserted through the wall of the vagina. The eggs are then fertilized *in vitro* (in a petri dish) using the husband's sperm. As the resulting embryos grow, single cells are extracted from each to determine their gender. Only embryos of the "correct" gender are inserted into the woman. The other embryos are destroyed or "stored" (cryogenically) for possible later use. (In reality, these stored embryos are rarely used, giving rise to ethical concerns.) To help ensure success, two embryos are often implanted with the idea that if one doesn't survive, the other might. If a successful pregnancy ensues, the chances of getting the gender desired is, in fact, virtually 100 percent.

So what's wrong with that? you are asking. Well, apart from the ethical issues (discussed in more detail below), you need to consider these facts: Among younger women, IVF results in a successful pregnancy only 43 percent of the time per attempt. So 100 percent gets knocked down to 43 percent at the outset. Of course, you can opt for additional attempts. But be aware that each of these is a rigorous exercise, requiring careful priming and management of your cycle. This, in turn, requires a lot of hormonal treatment, which carries some risk and often is accompanied by unpleasant side effects, such as weight gain, bloating, swelling, and even blurred vision. And as you get older, the success rate declines below the 43 percent, and the hormonal treatments may put even more stress on your body and mood. Removal of eggs can also be painful and, like all invasive procedures, can carry risk.

In addition, you may get more than you bargained for. If you are successful, you may get multiple births. Some 38 percent of offspring resulting from IVF are twins. And some researchers have questioned the overall benefits of PGD, as well as its safety, pointing out that there is still inadequate data to support its widespread use for something like sex selection.

And, finally, there are the costs. A single attempt at sex selection using PGD and IVF costs about $18,000 to $20,000.

Most reputable fertility clinics will not use PGD for sex-selection purposes. They reserve it for "medically indicated" purposes, as, for example, in the case of couples who have a history of genetic diseases or multiple miscarriages. Often they restrict the use of it even in these cases to women close to forty or over.

The use of PGD for gender selection is illegal in many countries. It is not illegal in the United States, but its use for this purpose is condemned by most U.S. medical authorities. Even one of its original developers, Dr. Mark Hughes, rejects its use for sex selection, asserting that he helped create the method to treat and avoid disease and disorder and for that purpose only. Dr. Kevin Winslow of the Florida Institute for Reproductive Medicine sums up the concerns of many when he says, "It probably isn't an ethically good thing to do because you're creating embryos when you know you won't use half of them."

The Diet Sex-Selection Method

Claims have been made that diet can be manipulated to determine the sex of offspring. These claims, in fact, were the subject of a 1982 book called *The Preconception Gender Diet* by Sally Langendoen, R.N., and William Proctor. Of course, there is nothing new about such claims; the idea that diet can influence gender has been around for centuries. What *is* new is that *this* diet theory appears at least to have *some* scientific underpinning. In short, there *may* be something to it. But read on before you get too excited.

The new work is based on the research of Dr. Joseph Stolkowski of Paris and Dr. Jacques Lorrain of Montreal. Dr. Stolkowski noted that, in the 1930s, Dr. Curt Herbst of Germany had discovered that the sex of marine worms could be influenced by mineral manipulations. He decided to see if the same thing might be true in higher animal forms, and reported that it was. Diets high in salt, he claimed, resulted in significantly more bulls than heifers. Dr. Lorrain subsequently entered into a collaboration with Dr. Stolkowski, to see if diet could similarly influence *human* sex ratios.

The first results of this collaboration were reported in the *International Journal of Gynaecology and Obstetrics* in 1980. The two doctors put 281 women who wanted children of specific sexes on two different diets. The "girl diet" is high in calcium and low in salt and potassium. The "boy diet" is high in salt and potassium and low in calcium and magnesium. Husbands were told to go on the same diets as their wives—mainly in order to lend psychological support and help ensure that the wives didn't "fall off" the diets. The couples were instructed to stay on the diets for a month to six weeks before attempting to conceive. If they had not conceived within six months, they were told to discontinue the diets. Some 21 women dropped out of the study because they couldn't tolerate the diets or had negative reactions to them or because they lapsed and became pregnant before they had given the diets an adequate chance. The doctors reported that of the 260 women who stuck with the diets, approximately 80 percent conceived children of the desired sex.

No convincing hypothesis has yet been put forward to explain why this method works, if in fact it does. Researchers have speculated on a number of possible modes of action. Possibly the diets affect the secretions within the female reproductive tract in such a way that, depending on which diet is used, either X or Y sperm are better able to move through those secretions. Or perhaps the manipulation of the minerals in the diets affects the membrane of the egg, making it easier for one or the other type of sperm to penetrate. Perhaps the diets affect the female immune system, selectively favoring one type of sperm over the other, again depending on which diet is used. And possibly the diets affect the sperm cells, too, since the husbands also adhere to the diets.

Before you jump to the conclusion that you can eat your way to the baby of your dreams, be aware that there are some potential problems with the sex-selection method based on these diets. First of all, at this point, the method rests entirely on the work of Stolkowski and Lorrain, who have conducted but one clinical trial, and that trial involved a relatively small number of women. Independent confirmations have yet to materialize. Second, following these diets for even a few months may turn out to be unpleasant and highly inconvenient. Third, there is the possibility of health risk.

All who read *The Preconception Gender Diet* will note the multiple warnings it contains, cautioning readers to consult their doctors before going on these diets and then, once on them, to continue consulting their doctors. The reasons for this are apparent when you get into the specifics of the diets. The boy diet, in particular, may be risky for some people because it requires a higher salt intake than many doctors think is healthy. The authors also acknowledge that the amounts of calcium permitted by the boy diet "are significantly below what is regarded by most nutritionists to be the minimum daily requirement." The amount of magnesium permitted is similarly well below generally recommended levels.

The authors of the book argue, however, that those who stay on these diets for no longer than six months "probably" won't experience any difficulty. The trouble with this kind of reasoning is that it will take only a few isolated instances of illness attributed to these diets to bring the whole approach into disrepute. What worries us is that people who are already endangering their health with too much salt (a problem that is acute in the United States) will go overboard on this diet. It has too high a salt content to begin with. Some, thinking "if this much is a good thing if you want a boy, then more might even be better," may be tempted to add still more salt to an already salty diet. And some others, who do not become pregnant within six months, will inevitably be tempted to stay on the diet and "keep on trying." The authors point out that there are some people who shouldn't use the diets for even a month, given preexisting health problems; but will all those people heed these warnings or recognize that they fall into these at-risk categories?

Among the obvious "contraindications" to the boy diet is hypertension (high blood pressure or even a predisposition to hypertension). Drs. Stolkowski and Lorrain reported that four women who were trying for boys had to drop out of their study because the high salt intake resulted in edema, or water retention, a sign of potentially dangerous circulatory problems. Women on the girl diet have been warned to be on the lookout for kidney problems, too much calcium in their blood, excessive nervousness, and so forth.

Langendoen and Proctor concede that the Shettles method may have validity but argue that it is too complicated and takes too much of the spontaneity out of the sex act. We are willing to bet that the overwhelming majority of people who try both methods will decide the Shettles method is, by far, the easier to follow and is encumbered by neither the health risks nor the day-in-and-day-out hassle of a diet that is bound to become obnoxious to many people.

At the same time, we do not want to criticize this research too harshly. Unlike the Whelan method, which we discussed earlier, this one appears to be based on sound, scientific methodology. It is our hope that with further study the promising work of Drs. Stolkowski and Lorrain will not only be confirmed but will yield information that will allow for the development of dietary alterations less taxing and risky than those currently recommended by Langendoen and Proctor. If those things happen, it will be interesting to see what results might be obtained by *combining* all or various aspects of the Shettles method with those of a revised dietary method. Some Japanese researchers, in fact, are investigating these possibilities.

For now, however, we cannot recommend the dietary regimens proposed in *The Preconception Gender Diet*. And we must point out that the Shettles method yields the same results that have been reported for the diet method—and without the risks and inconvenience of the diets. Moreover, as previously documented, the Shettles method has been confirmed by numerous independent researchers throughout the world. The diet method hasn't been.

Part Three

How to Use the Shettles Method Successfully

First Order of Business:
How to Determine
the Time of Ovulation

Whether you want a boy or a girl, the most important thing you must do is figure out when you ovulate. The timing of intercourse relative to the time of ovulation is the most crucial factor in selecting either a boy or a girl. For a boy, you will time intercourse to coincide as closely as possible with ovulation. For a girl, you will time intercourse, at least to begin with, three full days before ovulation.

To review briefly what we stated in earlier chapters: Intercourse that occurs at or near the time of ovulation favors the smaller, faster, Y-bearing (male-producing) sperm, whereas intercourse that occurs well before ovulation favors the larger, hardier, X-bearing (female-producing) sperm, which are slower but are able to survive longer in the more acidic, preovulatory environment. There are other factors involved, which we discuss in the next two chapters, but timing is the essence of sex selection.

Test kits are available, without prescription, that measure the level of a particular hormone excreted in urine, a hormone that signals the onset of events leading to ovulation. We have a lot to say in this chapter about these ovulation-prediction kits, many of which can be quite accurate if used properly.

We hasten to add, however, that these kits are not the "magic bullet" that takes all the guesswork out of pinpointing ovulation time. We've received a lot of letters about these kits, and it is evident that many people are misinterpreting and otherwise misusing them. They were not designed specifically for sex selection but, rather, to aid people with family planning, infertility problems, and so on. They *can* be used in conjunction with sex selection, and in this chapter we tell you how to do that. At the outset, however, we want to say that the use of these kits is entirely optional. They are expensive, and in many, probably most, cases, they are not necessary.

The truth of the matter is, there still is no fail-safe method of determining with absolute certainty the time of ovulation in most women. Even laboratory blood tests and ultrasound monitoring are not entirely accurate. In any case, it would be both impractical and very expensive to have these lab tests done on a regular basis. Nonetheless, the situation *is* better than ever before—and *most* women *will,* with the instructions we provide, be able to determine their ovulation time with improved accuracy.

With the technologies available for in-home use, even those women who have highly irregular cycles should now enjoy a greater sex-preselection success rate.

How the Menstrual Cycle Works

Before we proceed to the actual techniques of determining ovulation time, let's go over a few of the basics related to the dynamics of the menstrual cycle. The cycle is said to begin with the first day of bleeding. Bleeding often lasts from three to seven days, may vary from cycle to cycle, and is not the same for all women. About five days of bleeding appears to be average, though there may still be some slight "spotting" of blood after that. The bleeding occurs when the lining of the uterus is shed. This shedding occurs each month in normally fertile women, provided they have not, in the course of the preceding cycle, become pregnant. (There can be other reasons for not experiencing your period, and you should consult your doctor if you miss your period.)

Once the period is over, a structure ("follicle") in the ovary begins to ripen, preparing for the release of the egg that is contained within it. All these events are under the control of hormones, which are turned on and off at appropriate times by the hypothalamus and pituitary glands of the brain. Once the follicle begins ripening, it releases a hormone of its own—estrogen. It is the estrogen, in fact, that helps turn the bleeding off and promotes healing of the uterine lining. Estrogen helps get the womb ready for another attempt at pregnancy. It also makes the cervical secretions more hospitable to sperm as ovulation approaches. The first half of the menstrual cycle, dominated as it is by the ovarian follicle and the estrogen it produces, is often called the follicular phase of the cycle.

Ovulation itself—the actual release of the egg from the follicle—is stimulated by another pituitary hormone called the luteinizing hormone (LH). Other hormones have "ripened" the egg and brought it to the surface of the ovary, where it bulges outward in its follicle. LH, which surges some hours before ovulation occurs, gives the egg the extra "kick" it needs to burst from the follicle and pass into the fallopian tube, where, if pregnancy is to occur, it must be fertilized by sperm. (It is the LH, by the way, that the ovulation-prediction kits detect in urine samples, which are tested regularly during the fertile period of the cycle. By being able to tell when the LH surge begins, it is also possible, in many cases, to predict when ovulation is most likely to occur.) After the LH surge, the peaking of estrogen, and ovulation have occurred, the now-eggless follicle assumes a new identity as the corpus luteum, or "yellow body," which is what it resembles. The corpus luteum is a structure on the surface of the ovary that begins secreting a new hormone right after ovulation takes place. That hormone is progesterone.

The second half of the cycle—the postovulatory part—is called the luteal phase of the cycle, after the corpus luteum and the progesterone that it produces. Progesterone causes rapid development of the endometrial lining of the womb, such that it will be able to hold and nourish a fertilized egg, should one show up. The progesterone has a number of effects. One of these—particularly important to would-be sex selectors—is that it abruptly changes the nature of the cervical se-cretions, making it more difficult for sperm to penetrate into the female reproductive tract. If no egg is fertilized in the course of the menstrual cycle, then the corpus luteum "shuts down" and withers away; the hor-mones that have been preparing and maintaining the lining of the womb for possible pregnancy are turned off. The lining of the womb begins to degenerate and shed. It is ejected via the monthly bleeding "period." One cycle thus ends and a new one begins.

The *average* cycle is twenty-eight days, beginning with the first day of bleeding and ending on the day before bleeding begins anew at the start of the next cycle. The average day of ovulation is day 14, midway through the cycle. But please note that we are using the word *average* here. Most women *do* ovulate close to the middle of their cycles, but there is still considerable variability from woman to woman, and even within individual women, from month to month.

A few women have relentlessly irregular cycles, the length of which cannot be predicted from month to month. Some others have consis-tently short cycles or consistently long ones. Cycles of twenty-four days occur, as do cycles of forty days. Generally, though, there is more regu-larity than irregularity, and most women *can* discern patterns. Even those with highly regular cycles, however, must be aware that various forms of physical and psychological stress, illness, change of eating or sleeping patterns, smoking, heavier-than-normal intake of alcohol, and so on can result in transitory cycle irregularity.

The good news is that even many of those with irregular cycles can still determine the probable time of ovulation within individual cycles. The complex biochemical events that occur during the cycle are dra-matic enough to manifest themselves in physiological changes that can be detected, often with considerable accuracy, right in the home.

Let us now review the different methods of determining time of ovu-lation. We start with the cervical mucus (CM) method, which, in a *majority* of women, still provides the most useful and predictive infor-mation at the lowest cost. Then we talk about basal body temperature (BBT) charting, which, when used properly, can also be very helpful, especially in women with regular or fairly regular cycles. We recommend that women use both methods during the "test" cycles in which they try to determine their time of ovulation.

After we've discussed these two methods, we examine the ovulation-prediction kits, which can be used in conjunction with CM and BBT to provide a third source of information. We'll tell you how to coordinate all of these methods. And if one method doesn't work for you, then another is likely to do so.

Some may wonder why we don't *start* with the ovulation-prediction kits since they are the newest development, are based on the latest technology, and claim such high success rates (90 to 100 percent). The reasons that we continue to emphasize CM and BBT are many. Among other things, these kits can be expensive, although the prices are coming down into the $15-to-$40-per-month range. In addition, the success rates claimed for them relate merely to success in identifying the most fertile period of the menstrual cycle. Since that can still cover two to three days in many cases, the precision of these tests, especially as they are generally used, is not always sufficient to yield the desired result in sex selection. We'll tell you how to use the tests properly if you opt to use them, and we identify those women most likely to benefit from them.

For now, let's begin with the tried and true. Remember, the high success rate we've reported over the years has been obtained using, first, only the BBT and then, in more recent years, the CM method in conjunction with BBT charting.

The Cervical Mucus Method

For many years, researchers including Dr. Shettles thought that the "temperature method" of charting the cycle and determining ovulation time was the most reliable. The temperature method is still useful and will be described in more detail shortly, but it does have certain disadvantages. A better method, one that has been proving itself remarkably effective as a means of natural fertility control (both to avoid and achieve pregnancy, depending on one's goals), relies on observations of the *cervical mucus* at various times during the monthly cycle.

We have previously in this book only briefly mentioned the cervical mucus (CM) method. We have pointed out that the CM, which is secreted by cells in the cervix, is thicker, cloudier, and less abundant early in the cycle; as ovulation approaches, it becomes thinner, more transparent, and usually, but not always, more abundant. It is thinnest and reaches its maximum elasticity (stretching easily) at the time of ovulation, when it assumes the consistency and appearance of raw egg white. It can be observed on tissue, without need of internal examination, and, as we will see, can be a highly reliable means of assessing where you are in your cycle.

Observations of the CM are of central importance in a form of natural birth and fertility control called the ovulation method. The ovulation method is based largely on the work of two highly regarded Australian medical researchers, Dr. Evelyn Billings and Dr. John Billings. The method, in fact, is often referred to as the Billings method. It is based on recognition of "fertile" mucus and avoidance of intercourse on days when fertile mucus (which allows for easy penetration by sperm) is present. The ovulation method should not be confused with the so-called rhythm method, which is based on calendar calculations of average cycles and is notoriously unreliable as a means of birth control. The ovulation method is based on physiological observations that tell you what is actually happening in the reproductive cycle at any given time.

The ovulation method has become increasingly popular, especially in the last couple of years, as a means not only of birth control but of overcoming some forms of infertility. Many previously infertile couples have achieved pregnancy after studying the method, for it helps them to identify those days on which they have the best chances of conceiving.

Several studies have indicated that the ovulation method can be an extremely effective means of birth control. An international study, involving women in five different areas of the world, was carried out. Many of the women in this study were illiterate but were still able to master the technique, laying to rest the criticism that the ovulation method would be effective only in industrialized nations. As a matter of fact, a World Health Organization five-nation study showed that the *highest* success rates were achieved in the least developed nations, possibly due to greater motivation caused by pressing overpopulation. As the International Planned Parenthood Federation pointed out (*Research in Reproduction,* July 1982): "It is important to note that the method was relatively successful in all five of the countries, of which three were developing nations."

Various studies indicate that, when used properly, the ovulation method of birth control can achieve a 98.5 percent success rate. Of course, there is a difference between using it as a method of birth control and using it as a method of sex selection. If you plan to use it for birth control, you should first take instruction in the method from a teacher certified in the technique. Classes in the ovulation method are available in all metropolitan areas and, increasingly, in smaller communities, as well. Women's centers, student health services at college campuses, and various other groups and organizations may have classes or information on where you can get instruction. You may also want to write to one of the following groups for information on possible classes in your area:

Billings Ovulation Method Association—USA
P.O. Box 16206
St. Paul, MN 55116
Telephone: (651) 699-8139
info@boma-usa.org
www.boma-usa.org
www.woomb.com

The Couple to Couple League International, Inc.
P.O. Box 111184
Cincinnati, OH 45211
Telephone: (513) 471-2000
Toll Free: (800) 745-8252
www.ccli.org

In England:

> National Association of Ovulation Method Instructors
> The Billings Method Centre
> c/o 4, Southgate Drive
> Crawley
> W. Sussex RH10 6RP
> England
> www.billingsnaomi.org

It is not necessary to take classes to use CM observations for sex selection, though it certainly wouldn't hurt. The success rates we have cited for the Shettles method are not dependent on taking such classes. Those who have very irregular cycles and cannot get a "fix" on their ovulation time, using the information contained in this book, can no doubt improve their chances by taking classes. Or if you are interested in natural birth control or have been having difficulty getting pregnant at all, then, by all means, take classes if you can. Dr. Shettles never hesitates to recommend them, since they can be used for so many things: birth control, overcoming some forms of infertility, *and* aiding in sex selection.

An intermediate step between relying on the instructions in this book and taking actual classes from a certified instructor would be to study one of the books that have been written about the ovulation method.

Most couples will find the instructions we provide below adequate.

CM Basics

The first thing to be aware of is that estrogen makes the cervical secretions thinner, clearer, and usually more copious the closer you get to ovulation. This is nature's way of making it easier for the sperm to get to the egg. Thicker, cloudier mucus is more difficult to penetrate. What you must first start doing is to follow the progress of your CM over a number of "practice" cycles. That way you will get the hang of what the mucus looks like at different stages of the cycle; you are likely to discern patterns, especially if you have fairly regular cycles.

You will notice that when the CM first appears, there will not likely be much of it and it will not be particularly runny or elastic. As the cycle progresses, you will observe that the CM becomes clearer and will stretch farther between your fingers without breaking. Finally, you will find that it abruptly changes back, from a highly fluid, clear state to a cloudy thick state, indicating that ovulation has, in all likelihood, already occurred. (Some think ovulation occurs just after this change, but it seems more likely that ovulation precedes the change. In any event, the abrupt change in the nature of the secretions tells you that you are within, plus or minus, a few hours of ovulation.)

How do you find the CM to examine it in the first place? You've probably noticed it on many occasions already, but, in any case, it is usually not necessary to swab the interior of the vagina to retrieve it. Simply wipe the outer portion of the vagina with a clean tissue right after urinating. The mucus will be evident on the tissue if you are producing any at that stage in the cycle. During your practice cycles you may want to keep your CM charts, about which we'll say more shortly, in the bathroom so that you can make notations after each urination.

We have heard from women who say they can't find any CM on tissue and have to "probe" for it with a finger. If you are one of those women, here's the proper technique for making a CM "probe": Place your right foot on a chair or stool. Lean forward a little with your right elbow resting on your leg just above your knee. With your left hand press on your abdomen just above the pubis. This depresses the uterus and pushes the cervix forward, making it easier to reach. Then insert the index and middle fingers of your right hand into the vagina and gently touch the cervix, which will feel a little like the tip of your nose. (Make sure you have washed your hands carefully—and no long fingernails *please*.) If there is CM "at hand," you should be able to feel it and retrieve some for observation.

In most cases, however, as we've said, the "probe" isn't necessary. Generally you will be able to see the CM on the tissue after urination, especially as you approach ovulation and it becomes thinner and more copious.

During these practice cycles you will, of course, have to use some form of birth control. You can't use the Pill because it interferes with the CM, and, in any event, for safety reasons, you will want to be off the Pill for several months before trying to become pregnant. Contraceptive foams and jellies may also interfere with the secretions and your interpretation of them. Condoms are the ideal form of birth control to use during sex-selection practice cycles. They don't alter the secretions, and, by keeping the vagina free of seminal fluid (which women inexperienced in the method may confuse with CM), they make it easier to learn how to interpret the secretions.

This pattern is typical (but, remember, yours may vary a little or even a lot):

Days 1 through 5—bleeding. No need to check for CM on these days. A few women do have CM during the bleeding phase of the cycle, but such women are very rare.

Days 6 through 8—no bleeding and no CM. These are often referred to as dry days. *Usually* women experience a few of these dry days after bleeding and before the CM begins to show up.

Day 9—the first mucus appears. This is the first "wet day," but it's not very wet. After wiping the vulva with a tissue following urination, you notice a small amount of CM. The main things to pay attention to are the color and consistency, or "stretchability," of the mucus. The amount of CM is less important. Usually, but not always, there is less in the beginning and more as ovulation approaches. Go more by quality than quantity. Once you have spotted the CM, keep checking carefully after *each* urination. On the first day that it appears, it will most likely be quite cloudy and thick.

Day 10—mucus is present. It is still thick and cloudy or creamy in color. Try the "stretch test": Put some of the CM between your thumb and forefinger. Now slowly separate the two. At this point the CM probably will not stretch very far before it breaks, leaving one glob clinging to your finger and another to your thumb. Make a note of about how far the CM stretches before breaking so that you can make comparisons later. Early on, it often doesn't stretch more than an inch or an inch and a half before breaking.

Day 11—CM is present and now is thinner and more stretchable. You may notice this simply by looking at the mucus on the tissue. There may also be more than on previous days. It is still opaque but not as cloudy as before. It has a wetter feel and a runnier consistency. It stretches farther between thumb and finger. (You can also test stretchability by letting some of the CM hang off the edge of the tissue, noting how far it stretches before it stops or breaks and falls.)

Day 12—CM is present and still more watery than on the previous day. It is also clearer than it was on the previous day. It also stretches a bit farther.

Day 13—CM is present and now *very* watery. The mucus now has the consistency of raw egg white. It is very clear and slippery. It is highly elastic now and stretches many inches without breaking. This looks like what people who follow the ovulation method call the peak day, when estrogen is at its peak and the CM is maximally runny, clear, and elastic. The secretions may now be copious. You might guess that this is the day of ovulation, but this is a practice cycle, so let's keep going and see if you would be correct. If you are—and this *is* ovulation day—then the secretions should soon turn cloudy, thicker, and inelastic again, as progesterone comes into the system, overriding some of the estrogen effects that prevail right up to ovulation.

Day 14—CM is present and is *still* clear and slippery. These conditions may continue for most or all of the day. What we learn from this, then, is that the "peak" symptoms can *sometimes* persist for more than one day. In many women, perhaps the majority, they *do* last only one day; in other women, however, they may persist for two days. (And a few women have peak symptoms for only a few hours.) In this case, let us say that the secretions are still "peak" throughout the morning of this day of the cycle. By midafternoon, however, they have abruptly changed so that they are now thick and cloudy again—enough so that you can immediately tell the difference. The estrogen-dominated part of the cycle has ended, ovulation has occurred or is occurring or will occur very shortly, and the progesterone-dominated part of the cycle is beginning.

By going through a number of practice cycles—at least three but, in any event, enough to make you feel confident that you are pinning down not the hour, perhaps, but at least the day of ovulation—you will get a "feel" for the CM method and, in all likelihood, learn enough about your cycle so that you can achieve your goal. You may well find that you follow a fairly regular pattern and that your peak symptoms, for example, last a reasonably predictable period of time. Even in the absence of persistent patterns, however, often you can still discover your day of ovulation.

The CM Chart

It is of vital importance to keep charts. You can use graph paper or simply take a sheet of plain paper and rule it off into little boxes so that it looks something like Figure A. Put the date at the top, as indicated in the example. Use the month in which the cycle begins. Put the days of the month or months in the row of boxes beneath the cycle days. Don't confuse the two. Note that in Figure A the first day of the cycle is April 12 and the last day is May 9. This particular cycle lasts twenty-eight days.

This chart is based on the example given earlier in this chapter. The woman, as you will recall, bled between days 1 and 5. Therefore she has placed a *B* (for "bleeding") in each of the first five squares. On days 6 through 8, she experienced neither bleeding nor wetness in the form of CM. She has therefore placed the letter *D* (for "dry") in the boxes that correspond to those days. On days 9 through 14, the woman observes the CM in varying degrees of clearness, wetness, and stretchability. She places the letter *M* (for "mucus") in the boxes that correspond to those days. In addition, she puts a circle or *O* (for "ovulation") around the *M* in the box that corresponds with the fourteenth day, as this was when the CM truly "peaked" and ovulation likely occurred.

You will note that the woman has also put an *M* in the boxes corresponding to days 15 and 16, because she noted some CM on those days, too, though it was of the thicker, cloudier type that appeared earlier in the cycle. Most women continue to notice CM for two or three

Name								Year 2007		Month June			Doctor						
Cycle Day	1	2	3	4	5	6	7	8	9	10	11	12	13	14	15	16	17	18	19	20
Day of Month	12	13	14	15	16	17	18	19	20	21	22	23	24	25	26	27	28	29	30	1
	B	B	B	B	B	D	D	D	M	M	M	M	M	(M)	M	M	D	D	D	D

21	22	23	24	25	26	27	28	29	30	31	32	33	34	35	36	37	38	39	40
2	3	4	5	6	7	8	9												
D	D	D	D	D	D	D	D	D											

Figure A

or even more days after the "peak." After that, the woman in the example notes only dry days until bleeding begins again and so marks each box accordingly.

You should, for your practice cycles, keep checking the CM every day throughout the entire cycle. Once in a while some mucus will show up all of a sudden even late in the cycle, but at that point, it rarely is anything other than "infertile" mucus of the thick, cloudy type. Be particularly conscientious about checking for the mucus as soon as your bleeding or major bleeding ends. (You might have slight "spotting" after the period is largely over, check for CM even on those days.) Check your CM after each urination and note its color and consistency. Keep notes on the quality and quantity of your CM *daily*. You can put these notes right on the same sheet of paper with your chart so that everything relevant to a particular cycle will be in one place.

Thus, if you were the woman whose sample CM chart we've been discussing, you would write under the boxes the day on which mucus first appeared something along these lines:

> *Day 9—Mucus first appeared for first time today at about 10:30 A.M. It was thick and cloudy and would barely stretch at all; there was only a little on the tissue. It remained the same each time I checked throughout the day.*

Continue to make such notes for each day on which mucus appears, even after you think ovulation has occurred. You will find, once you get started, that this takes only a few minutes each day, and the records you create in the process will be of invaluable help to you when you make the actual try for a boy or girl. Make your final notations at the end of each day. Thus a day that starts out "dry" but ends up "wet" should be indicated, on the chart, with an *M*. Your additional notes will tell you what happened throughout the day.

We must emphasize the fact that different women will exhibit different CM patterns. And even an individual woman is likely to see at least some differences from month to month. Some women will go directly from their bleeding days to the "wet" days, with no "dry" days in between. Others may have prolonged "dry" days and only one or two "wet" days. Whatever the case, the CM method will give you important clues as to what is happening in your cycle at any given time. Even if your cycle is highly irregular, the quality of your CM can tell you whether you are potentially fertile on any given day.

Bear in mind that illnesses, a lot of tension or stress, or sudden changes in your environment, diet, eating and drinking habits, and so on can alter your CM pattern. So pay special attention to the secretions when anything in your routine is markedly altered. Be aware, too, that breast-feeding and many forms of infertility usually result in scanty or absent CM. Don't jump to the conclusion that you are infertile, however, just because you have little or no CM for a month or two. Generally, physicians don't regard couples as infertile until they have failed to conceive for at least a year. If you feel that your CM is abnormal, however, that is certainly worth reporting to your doctor.

Some women, especially those who have particularly long cycles, *may* have a day or two, or even several days, when CM is present, followed by one, two, or more "dry" days and then another set of "wet" days. Such irregularities make the utility of charting several practice cycles evident. When irregularities persist, you must be especially careful to be guided by the quality of the CM in order to determine which is your "peak" day, the probable day of ovulation. Those women who have particularly irregular cycles will almost certainly want to use *additional* methods of timing ovulation, along with the CM method.

A good many women have no trouble at all in finding their "peak" day. Some report that they become so expert at this that they can tell when they are "peaking" without so much as a visual check of their secretions. Some say they can tell by the feeling of wetness in the vulva area. We do recommend, however, that you *look*. And, in any event, you must remember that the peak can last for more than a day. When you think your symptoms are as "peak" as they will get, check them as frequently as possible.

Remember, too, that you cannot use averages to plan your own individual sex-selection strategy. Nonetheless, knowing what the averages are can be of some help, especially in the beginning, in terms of knowing what to begin to look for. With those words of caution, we will tell you that various studies have shown that 5.9 days, *on average*, elapse between the first sign of CM and the time of ovulation. *Your* personal average may turn out to be quite different, but taking all women together and averaging the results, that is the approximate interval between the first show of CM and ovulation—just about six days. This tells you, when you start out, not to jump the gun and assume at the first sign of CM that you will soon be ovulating. Be patient. Study your cycle. You have time.

The BBT Method

We recommend that you use basal body temperature (BBT) charting with CM charting to gain as much information as possible. If there is consistent disagreement in terms of results obtained with the two methods, then we recommend that you rely on CM rather than BBT. Or you may want a third opinion in the case of persistent disagreement, in which case you should consider the ovulation-prediction methods discussed later in this book.

It is not yet time to discard BBT. It has worked very well for many women. We strongly urge you to give it a good try. There are some women whose CM is itself so unpredictable that BBT is the only practical option.

The temperature method is based on the widely confirmed observation that body temperature shifts abruptly upward about the time ovulation occurs. The reason for this shift is that estrogen, which holds

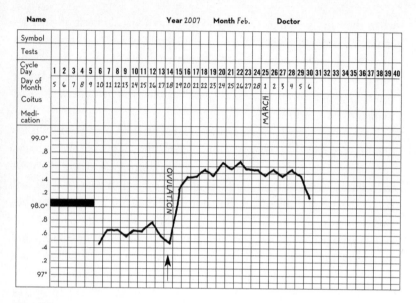

Figure B

temperature down, gives way to progesterone at the time of ovulation. Progesterone has the effect of raising the temperature.

Your task, once again, will be to keep a daily chart—this time of your temperature. A typical BBT chart will look like Figure B. You can make your own chart with a ruler and a plain sheet of paper, or buy some graph paper. Put the year and the month on each chart. The chart should show the temperature along the left side, between 97 and 99 degrees, with each box in between representing a tenth of a degree (97.1, 97.2, 97.3, etc.). In Figure B the cycle began in February. The first day of bleeding, which marks the first day of the cycle, was on February 5. Therefore, the number 5 is placed in the box under cycle day number 1 and so on. In this example, the cycle lasts thirty days, ending on March 5. (You can disregard the other boxes, for coitus, medication, etc., shown in the sample.) There is no need to take your temperature during bleeding days. Simply mark those on your chart, as shown in the example, by blacking out the box that corresponds to 98.0 degrees, which is average for women. Bleeding, in the example, lasted five days, so the first five days of the cycle are blacked out.

On day 6 of the cycle, the woman in the example noted no bleeding and so began taking her basal body temperature. For this you will need a BBT, or "ovulation," thermometer. You can obtain one without prescription at nearly all drugstores. They are generally inexpensive. Most have calibrations from 96 or 97 degrees to 99 or 100 degrees, with easy-to-read markings for each one-tenth of a degree in between.

This temperature is called *basal* body temperature because it is a measure of your body's *base* temperature, that is, the temperature as it is when you are at rest and under as little stress as possible. Nearly everything you do can affect your temperature, raising or lowering it above or below the true baseline. Therefore—and this is very important—you *must* take your BBT the first thing each morning *before* you get up, walk about, eat, smoke, have sex, and so on.

Keep the thermometer next to your bed and take your temperature first thing each morning, right after you wake up. (Instructions come along with the thermometer. Be sure to follow these with respect to the proper handling, cleaning, and storage of the thermometer.) Put the thermometer under your tongue for exactly three minutes. Have a watch or clock handy so that you can be precise about this—and keep your hands off the thermometer while it is in your mouth. At the end of the three minutes, note the temperature carefully and record it on your chart by placing a dot in the appropriate box. Repeat this first thing each morning. Connect the dots with straight lines, as shown in Figure B.

During your practice cycles, which will allow you to become familiar with your temperature pattern, try to maintain a fairly regular regimen with respect to diet, amount of sleep you are getting, exercise, and so on. *Don't* try to familiarize yourself with your cycle while on a trip or during the middle of a move, a change in jobs, or the like. A regular schedule will make it much easier for you to get the most accurate temperature profile of your menstrual cycle possible.

Be aware that physical and emotional stress, illnesses, including even colds, smoking, eating and drinking more than usual, sleeping poorly, and so on can all affect your BBT. Despite the occurrence of any of these things, however, keep on taking your temperature. Before long you will begin to see patterns on your chart. Eventually you will feel confident that you have gathered enough information, during periods when you were relatively free of stress and disruption, to make an adequate assessment of your ovulation time.

The thing you will be watching for, while BBT charting, is the sudden upward temperature shift that occurs daily until you start bleeding again. That will mark the beginning of a new cycle, at which point you should begin a new chart, as well. Almost all women who keep accurate charts find that their BBTs are noticeably lower before ovulation than they are after ovulation, reflecting the two overall phases of the cycle. Temperatures prior to ovulation are *usually*, but not always, somewhere between 97.4 and 97.8 degrees, a fairly narrow range. After ovulation, the temperature increases, *usually* by four-tenths to a full degree in a single day. Then it tends to stay high—almost always above 98 degrees, though it can still dip up and down a bit. In the postovulatory phase of the cycle, the BBT typically lingers between 98.2 and 98.6 degrees. Then, as the cycle nears its end and new bleeding approaches, the temperature drops sharply. As with CM charting, it is useful to keep notes on your activities, which later on may help you account for various irregularities in your BBT charts.

If you look again at Figure B, you'll see that the overall pattern is typically biphasic, meaning the temperatures are lower during the preovulatory, estrogen phase of the cycle and higher during the postovulatory, progesterone phase of the cycle. This woman started taking her temperature on day 6 of her cycle, recording a 97.4-degree reading. (Observe that the proper box for each reading is *above*, not below, the line indicated.) The readings continue pretty much in the same narrow temperature range between days 6 and 13. Then on day 14, this woman's temperature dipped a little below where it had been for the past several days. This often happens just before the sudden rise in temperature, which in this case occurs on day 15. (Some studies indicate that there is this small but noticeable dip just before the sharp rise in 75 percent of all women.)

When things are as clear-cut as they are in this case, and they often are, a woman can be fairly certain that she has found the "shift" in temperature and that ovulation has occurred, is occurring, or will occur within a few hours of that shift. Once the temperature goes up and stays high, you can be sure that ovulation has already taken place.

Not everyone agrees as to exactly when ovulation occurs in relation to the temperature shift. Some say it occurs right at the dip—at that last low temperature before the sharp rise; others say it occurs on the day of the abrupt rise. The best studies indicate that it occurs at the dip or soon after the rise *begins*. We are talking about a fairly short period of time, in any event. In the case of the woman whose chart is shown in Figure B, ovulation would most likely be occurring, if her temperature-charting is correct, at the dip on day 14 or early on day 15, or in between.

Not all charts are going to be as easily interpreted as the one seen in Figure B. Most will look something like that. But 10 or 15 percent of all women may show a much more gradual increase in temperature. The overall result will still be biphasic, but it may look more like Figure C. The woman who produces a chart that looks like this may find herself confused, at least initially. On day 10 of this cycle (June 11), there is a dip preceding a sharp rise. Is *that* the day of ovulation? It seems very unlikely that ovulation would occur that early in a twenty-nine-day cycle. So the woman is patient and keeps charting. The temperature does go up for the next two days but then it starts down toward a second dip, hitting bottom on day 14. The next day it is up again, though it's still under 98 degrees. The woman isn't sure what's happening, especially since on day 16 the temperature is down a little again, then rises once more the following day.

Only when the cycle is complete and the woman has charted several cycles, all of which look something like this, is it possible to get a handle on what is happening. In this case, the woman must ask herself: Did ovulation occur near the dip on day 14 of the cycle, at the peak on day 15, or at the dip on day 16? The woman in this case must pay particularly close attention to her other ovulation symptoms, especially the CM. This woman noted that her CM "peaked" closer to the dip on day 14 than to the rise on day 15 or the smaller dip on day 16. She there-

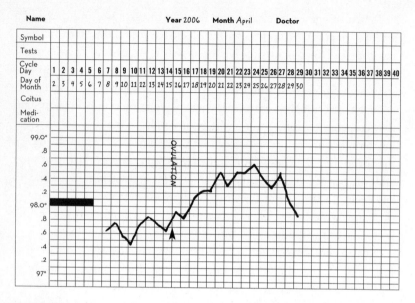

Figure C

fore concludes that her ovulation occurs at or just a bit after the dip on day 14.

Some women will find their charts even more difficult to interpret. They may be so erratic that no pattern whatever is evident. *Keep on charting.* If, however, after six months you are still at a complete loss, using both the BBT and the CM, it is possible that you have menstrual irregularities of a nature that only your doctor can properly assess. (Very painful menstruation, excessive bleeding, or noticeable bleeding in mid- or late cycle should always be called to the attention of your doctor.) Of course, rather than wait six months, some women for whom neither CM nor BBT seems to be working will want to try the ovulation-prediction kits to gain further insight into your ovulatory pattern.

Some women report that their temperature shift often amounts to *less* than four-tenths of a degree. For that reason they doubt that they have ovulated. Again, only by repeated charting can you determine whether this smaller shift does indeed signal ovulation in your case. If this small shift repeats and, especially, if it coincides with your "peak" CM symptoms, then it is likely that it *does* indicate ovulation.

Since you will most likely be using BBT with CM, it is worth mentioning what researchers have found with respect to the relationship of the two. Studies confirm what we have been saying. In the typical or "average" woman, the peak CM symptoms will be observed in the same twenty-four-hour period in which laboratory tests indicate ovulation occurs. In that same typical woman, the twenty-four-hour period between recording the last low temperature and noting the sustained rise in BBT has been shown to coincide with the day of ovulation. Thus everything seems to mesh, at least in the *average* case.

Kits to Predict Ovulation

There are several ovulation-prediction kits on the market. A fair number of people, unfortunately, have assumed that these kits actually were designed for sex selection and have jumped to the conclusion that the claimed 90 to 100 percent success rate for these tests meant they would have 90 to 100 percent success in conceiving the child they wanted. Worse, an equal number concluded that what the tests detect—the onset of the luteinizing hormone (LH) surge—is the same as ovulation. Thus they decided that ovulation was taking place at the first indication of the LH surge.

This is absolutely not true. The LH surge *precedes* ovulation by an amount of time that remains open to debate and varies to some extent from woman to woman. A couple trying for a boy who use one of these tests and mistakenly conclude that the LH surge means ovulation is occurring could easily end up with a girl.

This is not to say that these tests don't represent a real breakthrough. They do. But they were not designed specifically for sex selection, and if you are going to use them to help you preselect the sex of your next child, there are many things you need to know and do. Many people have figured out on their own how to use them—and did so successfully.

The point is, this is neither a time for great rejoicing nor a time of great despair. These tests are *not,* as we've said before, the long-awaited simple, easy-to-use, utterly reliable "magic bullet" that will make confusion over ovulation time a thing of the past. Used properly, they *can* help, but they are far from being the final answer.

Here's an excerpt from a letter sent to us by one of our readers. It is typical of a great many we received and points out some of the pitfalls people are encountering as they try to make sense of these tests.

> Dear Dr. Shettles:
>
> Realizing that timing is by far the most important feature to [your] method, and also being that we do not want to wait three months for practicing, I purchased an at-home ovulation-detection kit. They call it a *prediction* kit, but I believe it should more accurately be called a *detection* kit, as it really can only indicate that you *are* ovulating, and not *when* you *will* ovulate. In theory, this could only be used when trying for a boy [since] is cannot predict ovulation before it occurs. What do you think of these kits? Are they as accurate as CM and BBT?
>
> Thank you,
>
> Mrs. B.

This letter is full of misconceptions. First of all, these really are *prediction* tests, not detection tests. In theory, they can be used to help plan the timing for both the boy and the girl. In terms of their accuracy in sex selection, we don't have enough data yet to compare them with CM and BBT. To be used in sex selection, they must be used somewhat differently from the way they are typically recommended by their manufacturers.

Here are excerpts from two more letters that will call attention to the *central* difficulty people are having with these tests when it comes to applying them to sex selection.

> Dear Dr. Shettles:
>
> I have just finished reading your book. I found it very informative, but I would like an update on some of the mate-

rial. I'm having difficulty determining when I ovulate and hope that a test kit will help me determine this with greater accuracy. My question is this: The test states that ovulation can be predicted 20–44 hours in advance. If I'm trying for a girl, should I abstain from sex after the LH surge has been detected?
Sincerely,
Mrs. E.

Now compare this last letter with this one:

Dear Dr. Shettles:
I have purchased an ovulation-predictor kit. Could you tell me if this kit will help us even more? In the kit, they say, "A simple-to-read color change in your test results indicates this [LH] surge. Most women will ovulate within 12 to 24 hours after this surge." So do you recommend [for a boy] to have intercourse 12 or 24 hours after this surge? I feel like the more things I can try the better our chances will be.
Sincerely,
Mrs. C.

Both Mrs. E. and Mrs. C. correctly conclude that the LH surge is not the same as ovulation, and both are trying to make the information they've gathered from their test kits work—in one case for a girl, in the other for a boy. But notice that Mrs. E.'s test kit instruction booklet has advised her that ovulation will follow the surge by twenty to forty-four hours. Mrs. C.'s kit, on the other hand, states that ovulation will follow in twelve to twenty-four hours!

Both can't be right. Differences in the tests can't account for these discrepancies. There is only one LH surge in each cycle, and it is either detected or it isn't. These tests, in fact, are *very accurate* in detecting the LH surge. The problem is, the people who make these tests are inconsistent in their estimates of how long after the surge ovulation actually occurs. According to these two manufacturers, it could be anywhere from twelve to forty-four hours. That's quite a spread—far too wide a spread for sex-preselection comfort.

How do they get by with this kind of inconsistency? Well, first of all, even the world's leading experts disagree somewhat over this issue. But there's also confusion over terminology, and once it is sorted out, things appear clearer. Beyond that, however, manufacturers of these products are not concerned about sex selection; they simply have set out to provide women with a way of determining the most fertile portion of their cycles. Once you detect the LH surge, you know you are in a fertile period. Women who have been having trouble conceiving use these kits to detect their most fertile periods. They don't particularly care whether they have a boy or a girl; they just want to get pregnant.

Some people are probably also using these kits to help with natural birth control—a use all the manufacturers properly warn against. Fertility is not as great *before* the LH surge, but it is still great enough to result in pregnancy (more likely yielding female offspring) in a significant number of cases. By the time the LH surge is detected, a woman who has been having regular, unprotected intercourse may already have conceived.

We hope these letters have helped you to better understand some of the difficulties in using these test kits for sex selection. And perhaps by now you're already getting an inkling of how they might be used more effectively. The tests that our previous book called inaccurate were *not* these new tests. These are an entirely new breed. Let's find out more about them.

How the Tests Work and How to Use Them Properly

These tests are a product of what has been called the new biology, genetic engineering, recombinant DNA, and so on. Biologically engineered entities called monoclonal antibodies linked with enzymes produce color changes in specially prepared test strips, sticks, or tubes—in reaction to chemicals found in urine. The target chemical is the LH we've been talking about, the hormone that is released prior to ovulation—the hormone that, in fact, helps induce the ovary to release the egg. An abrupt, definite change in color in the test strip (going from a white or very pale color to a deep blue in most kits) signals the LH surge or lets you know that the LH surge has occurred.

Most of the manufacturers recommend testing once a day (usually first thing in the morning) for, in most cases, no more than six days in any given cycle. Each manufacturer has a formula that tells the user when to begin testing. The manufacturer typically says that once the user has detected the LH surge, it's not necessary to test further since the next two or three days define the most fertile period.

For use in sex selection, however, we advise you to use these kits somewhat differently, if you use them at all. And that, in fact, is the first thing you must decide: *Should* you use them?

If you can afford to, go ahead and use them as much as you want. They will give you additional useful information. But, again, that information will not be much better than what you can derive from the CM and BBT, especially if you have generally regular cycles and good CM production. If you fall into the latter category, you may want to use one of the ovulation-predictor kits through just one cycle, to see if it confirms your CM/BBT findings. Or if you have had a great deal of difficulty in determining your ovulation time using CM and BBT and/or have very irregular cycles, you may want to use a test kit for two or three cycles while you try to pinpoint your time of ovulation.

If you *do* use the kits, we recommend using one that permits you to test at any time of the day, not one that insists on testing only your first (early-morning) urine sample. Unlike the BBT, these ovulation tests are *not* affected by emotion, illness, or even most drugs. (Even the fertility drug Clomid doesn't affect them, although another major fertility drug—Pergonal—does.) So, unlike with the BBT, it is not important for you to test first thing in the morning in order to avoid misleading results.

Some tests insist on early-morning (first urine) testing because they are very sensitive to excess liquid intake. We like the Q Test system (which, like almost all of them, has an 800 number you can call for more information or assistance in using the product) because it allows you to test at any time of the day—provided only that you test your urine at *the same time* each day. Even with this test, however, you are advised to avoid excessive liquid intake before collecting your urine samples. We suggest that you not drink any liquids for a couple of hours before taking your urine samples. Excessive liquids may dilute the amount of LH in your urine and give false readings.

We don't like early-morning testing because of the nature of the LH surge. The LH "surge" is well named. It is a sudden, dramatic release of hormone, with an immediate, very dramatic rise in the concentration of LH in the blood—*usually* (but not always) occurring in the early morning. By the time the surge shows up in the urine, several hours have passed. One of the best studies we've seen on this subject indicates that a woman is most likely to detect the LH surge when she tests between 11 A.M. and 3 P.M.

So if you are going to test only once a day, pick a time between those hours and test each day at that *same time* during your testing period (which we'll get to shortly).

The same study indicated that the hours between 5 A.M. and 8 A.M. are actually the *worst* times to test. The second-best time is between 5 P.M. and 10 P.M.

If you plan to rely primarily on these tests to help you predict your ovulation time, we strongly recommend that you test twice a day, once between 11 A.M. and 3 P.M. and once between 5 P.M. and 10 P.M. Try to leave as much time between the two tests as possible, while staying within those time frames. Testing at noon and again at 10 P.M. would be an excellent choice, giving you a ten-hour spread.

The following scenario will help you understand why twice-daily testing is recommended if you plan to use these kits as the *primary* means of determining when you ovulate.

Mrs. H., following the instructions in her test kit, tests only once a day and always early in the morning, when she first gets up at 7 A.M. The test strip does not change color when she tests her urine on Sunday morning. But on Monday morning it does change color. Her instruction booklet tells her this means the LH surge has begun and that ovulation will follow in twenty to forty-four hours. Since she and her husband are trying to have a boy, they decide to schedule unprotected intercourse for twenty-four hours later, thinking that will place them close to ovulation.

In fact, this timing schedule might work in some cases, depending on when the LH surge really began. But let's suppose in this case that the surge actually started on Sunday morning. Since Mrs. H. tests early in the morning, the surge of LH could already have entered her blood but not yet shown up in her urine on Sunday morning. By testing early she missed detecting the surge that had already taken place. She and her husband assume that the surge actually began on Monday morning—but, in reality, ovulation could already have taken place by Monday morning. By waiting another day, Mr. and Mrs. H. place themselves in a gray zone, with their attempt to conceive occurring a day or more after ovulation has taken place. This gray zone attempt could result in no conception at all or, possibly, in a female conception.

Had this couple been testing at midday, they would have picked up an LH surge on the same day that it began. And if by chance the surge came at midday, their second test, in the evening, would certainly have picked it up.

You can imagine many other scenarios where things could go wrong as a result of once-a-day testing. The point is, when you test only once a day, the "surge" you detect may already be a day old. These test systems continue to indicate the LH surge for two or three days. They cease to do so only after ovulation has occurred and the progesterone phase of the cycle kicks in.

The tests usually are sufficiently predictive for sex-selection purposes (especially when used with CM and BBT) *if* you can detect the surge when it is no more than about half a day old (twelve to fourteen hours). If you test twice a day, your chance of catching the surge in its earliest stages improves enormously.

Despite all the conflicting figures on when ovulation occurs in relation to the surge, the best studies indicate that it takes place, on average, about thirty hours after the *onset* of the surge. Remember, however, that by the time you detect the surge in your urine—again assuming the ideal situation in which you are testing twice a day—the surge has already taken place five or six hours *earlier* in your blood. So, in general, once you detect the surge in your urine on the *twice-daily* testing schedule, you can assume ovulation will take place about twenty-four hours later.

Again, though, it is wise to continue to pay heed to your CM and BBT data at the same time. If you have very clear-cut CM and BBT information—that is, if your cycles are quite regular and CM and BBT tend to agree—it is unlikely that you will have any need for these ovulation-prediction tests to begin with. But if you do use them and get disagreement, we urge you to be cautious. If your CM data is very good, we'd go with that before we'd rely on a strongly conflicting test-kit prediction result. Better yet, try another practice cycle.

We'll give you some more specifics on how to use these kits in relation to specific attempts for a boy or a girl in the next two chapters. For now, we'd like to add that when you use the kits to try to discern your ovulatory patterns, do *not* stop using them after you've first detected the surge. During practice cycles (when you are not actually trying to get pregnant), you might test only once a day with these kits (ideally around midday) and continue to test even after you've detected the surge. See how long the surge continues to show up—and correlate your results with your CM and BBT findings. Since these tests are expensive, you won't want to start using them until you think you are within, say, three days of ovulation during your practice cycles. You can wait even longer if your other findings have already given you a pretty good idea of your probable ovulation day. You can stop testing as soon as the results indicate the surge is over and ovulation has taken place.

Remember that there are variations even in the most "predictable" menstrual pattern. Responses to these tests are going to vary from woman to woman and, in many cases, will vary somewhat within the same woman from month to month. In some months you may not ovulate at all, and thus you will not detect an LH surge. Be aware, too, that if you are approaching menopause, this may cause false readings due to persistently elevated LH levels.

Other Methods of Timing Ovulation

There is a method of determining when ovulation occurs that can be extremely precise. Unfortunately, not every woman can avail herself of this method, which is called *Mittelschmerz,* a German word referring to the middle of the menstrual cycle and a pain that is felt at the time of ovulation in the lower abdomen, usually on the right side. Some women feel the sharp twinge of *Mittelschmerz* at the exact moment the egg is ejected from its ovarian follicle. Some women also experience a *small* amount of bleeding at this time.

About 15 percent of all women are estimated to have this convenient and exact indicator of ovulation. If you think you are one of them, check to see if the midcycle twinge coincides with other indicators of ovulation. (There can be other causes for such pains, so don't automatically assume you are experiencing *Mittelschmerz*). Some women have midcycle pain related to ovulation but of a much less precise nature, with no definite sudden onset in the form of a "twinge." Such women may experience a dull ache during midcycle, an ache that persists for a day or longer. This may or may not be indicative of ovulation, so be sure to rely on other factors, as well.

The late Dr. Sophia Kleegman, a sex-selection pioneer, always informed her patients of the possibility of *Mittelschmerz* and, in fact, claimed that women who didn't ordinarily experience it could be *taught* to do so. She reported that fully 35 percent of her patients became aware of *Mittelschmerz* by practicing what she called the bounce test. Beginning on the ninth day of each menstrual cycle, Dr. Kleegman's patients were instructed to literally bounce, in a sitting position, on the hard surface of a wooden or metal chair, bench, or other unpadded surface. They were told to do this by sitting down abruptly two or three times in quick succession. This, Dr. Kleegman claimed, "brought out" or amplified the pain in many women, enabling them to feel it for the first time.

The women were instructed to write down the date when the pain was felt and then to see if it would occur at or about the same time in the next cycle. Dr. Kleegman did not believe that the bouncing actually caused ovulation to occur; rather, she thought that the pain, thus elicited, indicated ovulation had recently occurred or would soon occur.

In previous editions we've talked about something called Tes-Tape. Unlike the ovulation-prediction test kits we just discussed, Tes-Tape is inexpensive. It is available without prescription and may be used to help time ovulation, but only in conjunction with other methods. Tes-Tape was designed to help diabetics determine the amount of glucose (sugar) in their urine. It's a role of specially treated yellow paper that comes in a Scotch tape–type dispenser for easy use. The tape turns various shades of blue and green, depending on the amount of glucose that is present in the fluid. Many years ago Dr. Shettles and some of his colleagues discovered that glucose is also present in the cervical mucus and that the glucose content of the CM usually increases as ovulation approaches.

Thus it occurred to Dr. Shettles that Tes-Tape might be a useful additional aid in assessing CM and its relationship to ovulation. Some women use it with high accuracy, but others do not. Overall we do *not* regard it as highly reliable because it is difficult to use properly. If you decide to use it, tear off a three-inch strip of it and bend it over your index finger (you'll have to sacrifice that fingernail). Secure it to the finger with a small clean rubber band. Guide the finger gently into the vagina until the tip makes contact with the cervix. Try to go directly to the cervix, so that the Tes-Tape doesn't make contact with the more acidic secretions of the vaginal wall. (This is one of the major problems in using the Tes-Tape, as the acids will give you a false reading.) You will know you are touching the cervix, Dr. Shettles observes, when you make contact with something that feels like the tip of your nose.

When you do make contact (practice a few times before trying it with the Tes-Tape), hold the paper gently to the tip of the cervix for ten to fifteen seconds. No need to press hard. Then withdraw your finger quickly. Note the color of the tape where it made contact with the cervix. You begin this process after bleeding has stopped and continue once daily until you believe ovulation is at hand or has occurred. Do it in the morning, right *after* you take your temperature *(not before)*. In the early part of the cycle, the tape may not change at all, or, if it does, it will probably change to a light green. As you get closer to ovulation, however, the tape should become increasingly dark. Compare the colors to those you find on a color chart on the Tes-Tape dispenser. At the time of ovulation, the color of the tape should look like the darkest color on the chart (a deep greenish blue).

Again, however, remember that individual chemistries vary considerably. There is always the chance that the Tes-Tape will be in error in your case. So, to repeat, *don't* rely on it as your sole means of finding your ovulation time.

In conclusion, for the majority of women, we recommend the CM method most highly. We also urge you to use BBT charting in conjunction with the CM method. In cases of disagreement between the two approaches, rely more heavily on the CM method. If, for some reason, you find the CM method distasteful, go ahead and use BBT alone. With care, BBT can be made to be as accurate as CM. If you have *very* irregular cycles or want a "third opinion," try one of the ovulation-prediction tests, if you can bear the expense. If you are fortunate enough to experience *Mittelschmerz,* by all means make that an important part of your effort to pinpoint ovulation. Consider Tes-Tape as an additional aid, if you want to take advantage of as many "opinions" as possible.

Trying for the Boy—
What to Do

It is essential that you read this entire book and that you study the preceding chapter, in particular, with great care before making any sex-selection attempt.

We begin with the boy, rather than the girl, because the procedure for begetting male offspring is somewhat easier than the method for female offspring. You will recall that the intercourse that coincides as closely as possible with ovulation will be the most likely to yield sons. This has been shown to be the case in numerous studies involving *both* artificial insemination and natural insemination.

We don't pretend to know *all* the possible reasons why more sons result, given this timing, but it *has* been established that the boy-producing sperm are smaller and, under ideal conditions, are capable of swimming faster than female-producing sperm. Since the cervical secretions are most ideal at or near the time of ovulation, Dr. Shettles has long held that intercourse at that time should favor the male-producing sperm, which can use their greater speed to get to the egg first.

Even if intercourse takes place several hours before ovulation occurs, but within a day of it, the secretions still will be most conducive to Y-sperm penetration, permitting many of the boy producers to get into the fallopian tube to await the arrival of the egg. Some girl producers will arrive there, too, but the Y sperm should, for several hours at least, maintain a numerical superiority in the fertilizing zone. For the boy, then, you want to time intercourse for the day of ovulation.

It is easier to come in "on target" for the boy than it is to play the "waiting game" for the girl, as described in the next chapter. (Even those who are interested only in conceiving sons at this point should also read the girl chapter; by doing so you will learn more about what you *shouldn't* do.)

Time Is of the Essence

In the last chapter we told you how to determine your time of ovulation. That has to be your first and most important order of business. You will probably have to go through several cycles before you feel confident that you can tell when you are ovulating. If you have been on birth-control pills, you will want to wait *at least six months* after discontinuing them before trying to conceive a child of either sex. It has been shown, in numerous studies, that women who conceive within a few months of discontinuing the Pill are far more likely than other women to suffer miscarriages and other complications of pregnancy. Besides, once you've been on the Pill and have discontinued it, you will need time to get your cycle back to normal.

If you have been using an IUD, there is not the same danger, though Dr. Shettles and some others recommend that women wait a few months after removal of IUDs before attempting to conceive, just to make sure everything is back to normal. Contraceptive foams and jellies should *not* be used by couples who want to conceive boys. These substances are highly acidic and could thus be expected to most adversely affect the smaller, less hardy, male-producing sperm. And, in fact, studies have shown that pregnancies that result from failures of these methods of birth control yield a greater than expected number of girls.

The form of birth control that should be used during your practice sessions is the condom. This will not interfere with the acidity/alkalinity of the female reproductive tract and, in addition, will enable the woman to observe and test her cervical secretions on a daily basis, as outlined in the preceding chapter, without confusing seminal fluid with CM.

We recommend charting the CM as your first and best means of finding ovulation time. Remember that you may show a pattern of peak symptoms on two days in a row, rather than just one. That's one of the reasons why you should go through several practice cycles—so you won't "jump the gun." Even if you show a highly regular pattern, we suggest going through *three* cycles so that you can proceed with greater confidence. If your cycle turns out to be irregular, ultimately you may have to do a lot of cross-checking, using both CM and BBT.

We recommend that the BBT method be used in conjunction with the CM method. BBT will provide additional and, we hope, confirming information. If there is a conflict, place a higher trust in your CM readings unless you have some particular difficulty in interpreting them. If you are lucky enough to experience clear-cut *Mittelschmerz,* you can use that, as well, to help you pinpoint your ovulation time.

In previous books, where we have recommended primary reliance on BBT, Dr. Shettles has noted that, in the average woman, ovulation occurs somewhere between the dip in temperature and the sustained rise. If a woman took her temperature one morning and, from experience with previous charts, knew that by the next morning her temperature would be sharply up, then Dr. Shettles would recommend that intercourse be timed for that evening—about halfway between the dip and the top of the rise. If the woman was not sure that she had recorded her dip, then Dr. Shettles would recommend that she wait until the following morning (when the dip followed by the sharp rise would be evident) to have intercourse with the objective of conceiving a boy. The same recommendations still hold.

To repeat, if you are using BBT alone, intercourse for the boy should occur either on the day the dip is observed—if you are confident that this *is* the dip—or on the following morning, after the dip has been confirmed and the rise observed. In either case, you will be having intercourse during a twenty-four-hour period that we believe marks the day of ovulation. If you are confident that your morning temperature marks your dip, then intercourse later in the day would be ideal, rather than waiting until the next morning. The ovulation kits we discussed in the preceding chapter may also be helpful for those with consistently irregular cycles. (See discussion later in this chapter.)

Of course, we don't recommend that BBT be used alone. CM will give you more up-to-the-minute information. What we suggest is that you take your BBT before you get up each morning; then check your CM after each urination *throughout* the day (and even during the night if you get up to urinate). You will soon begin to notice correlations between the BBT and CM that will be very helpful to you in assessing the status of your cycle at any given time. You may find, for example, that at the point you record your last low temperature, you are very near ovulation (as indicated by the CM) and that you will have maximal chances of conceiving a boy if you have intercourse *that* morning rather than wait until the following morning. On the other hand, your CM readings may tell you that you still have plenty of time after that last low reading.

Your best chances of conceiving a boy are in the twelve hours before ovulation occurs. But you still have significantly higher chances of conceiving a boy than a girl when intercourse occurs within twenty-four hours before ovulation.

You may feel that you know when you are within twenty-four hours of ovulation but may feel less certain about knowing when it has already taken place. If you are relying on the BBT method, we don't recommend having intercourse for the boy any later than the morning on which you note that the temperature has risen abruptly. CM is even a better guide. When ovulation has occurred, your secretions will abruptly turn thicker and cloudier again. If you have been checking every few hours and suddenly note this change, we recommend that you *not* have intercourse. If you find, by charting several cycles, that your peak symptoms tend to last a day and a half, time intercourse near the end of that peak period—well into the peak period but before the change in CM takes place. Again, however, intercourse anytime during the last twenty-four hours of peak symptoms will increase your chances of having a boy.

Two Hypothetical Cases: Sally and Maria

Let's consider two cases, based on a composite of actual patients, to help clear up any ambiguities.

SALLY

Sally has been charting her cycle for three months now. She is happy to discover that her cycle is highly regular. Each has been twenty-nine days long. She is confident that she can determine her time of ovulation, and she and her husband feel they are now ready to try for a boy. Here's what they experience.

In her fourth cycle, the one in which they plan to make their actual sex-selection attempt, Sally ceases bleeding on the sixth day. On the seventh, she begins charting her BBT and CM. Her temperature follows a pattern much like it has in the past. So does her CM, which first shows up on day 10. The CM gradually becomes thinner and more copious until, on day 15, it peaks: It is as thin, clear, and stretchable as it ever will be. It has the consistency and appearance of raw egg white. At the same time, on the morning of day 15 of the cycle, Sally notes that her temperature, which has drifted slightly downward for the past few days, looks like it has reached its characteristic low point—the last dip before the rise. She is sure that by the morning of day 16 her temperature will be sharply up—at least half a degree. That is what has happened in each of her previously charted cycles.

Does she tell her husband to call his office and inform the boss he'll be a little late this morning? No. Another woman—with another pattern—might, but Sally knows from her previous charting that it is highly likely that her peak CM symptoms will persist all the way until the next morning. So she waits, wanting to get as close to the actual moment of ovulation as possible. When she takes her BBT the next morning she finds that, just as expected, her temperature has risen markedly. With some considerable excitement and a little anxiety, she quickly checks her CM. With relief she notes that, as in cycles past, it is still at its peak. She also knows, though, from past experience, that it won't remain this way much longer.

Sally tells her husband that ovulation is at hand. *Now* he can call his office. It's time to try for the boy without further delay.

MARIA

Maria is not quite so lucky as Sally. She has charted four cycles, and *each* has been a different length, ranging between twenty-seven and thirty-two days. Moreover, the most likely day of ovulation within each cycle, she and her husband have concluded, has been different. Nonetheless, they do feel that, *looking back,* they can tell just about when ovulation occurred in each cycle. Maria's charts are all biphasic, that is, they show the normal temperature shift. The trouble is, since the temperature shift occurred at a different time in each cycle charted so far, this shift cannot, in and of itself, enable Maria and her husband to predict the time of ovulation in any individual cycle. They know when it happened—but only *after it happened.*

Still, they feel they are ready to make an attempt for the boy in the fifth cycle. The reason for their confidence is the fact that Maria's CM readings have shown changes consistent with the BBT temperature shifts. This correlation gives them the confidence they need to proceed.

Here's what happens in the fifth cycle.

Maria stops bleeding on the fifth day. She has two dry days and then the CM appears. The CM has shown up on different days in each cycle, but once it starts it persists, getting thinner and wetter with each day up to the time of probable ovulation. There is always one day, in each cycle Maria has noted, on which the CM symptoms are more "peak" than on any other day. And looking back at the BBT charts for the previous practice cycles, Maria and her husband began to notice that those peak CM days occurred on the days of the last low temperature. This is what makes them feel that they have indeed found a way to determine ovulation time, despite Maria's considerable irregularity.

Therefore, in her fifth cycle Maria charts both her CM and her BBT but pays special attention to the CM. On the day she experiences her peak CM, she now *assumes,* on the basis of past experience, that she has recorded her last low temperature and that when she takes her BBT again the next morning, it will have risen markedly. She knows from past experience, too, that once the temperature rise has been recorded, her CM will already have turned thick and cloudy. Thus the day that she notes her peak CM symptoms is the day to try for a boy. She knows not to wait until the next morning. She and her husband have intercourse that day, and by evening Maria is glad that they did, for at that time she notes the sudden change in her secretions, from peak back to thick and cloudy. And, providing further confirmation that she did the right thing, the next morning Maria notes that her temperature has risen.

Use of the Ovulation-Prediction Kits

If you decide to use one of the ovulation-prediction kits (see discussion in the previous chapter) to help you with your timing for a boy, here's what we suggest:

First of all, if your cycles are regular or fairly regular but you simply want more assurance that you have pretty well pinned down your probable ovulation time, then by all means use one of these kits, if you don't mind the extra expense. Given a fairly regular cycle, you may want to use the kit during just one of your practice cycles.

You could, for example, start testing four days before your other calculations have, in previous test cycles, indicated that you would probably ovulate. For the first few days you might test just once a day—either between 11 A.M. and 3 P.M. or in the evening. On the day before you think you will ovulate you might test twice—once at around noon and again at around 10 P.M. If you pick up the LH surge while testing on this day, then your estimate of your ovulation time has probably been accurate: Ovulation will most likely occur on the following day.

What you should do then, however, is continue to test the following day, as well. Keep testing until the results indicate you are no longer in the LH phase. This will tell you that ovulation has occurred, though not precisely *when* it occurred. By this procedure, you should be able to gain valuable additional information about your cycle.

If you have consistently irregular cycles so unpredictable that the CM and BBT methods haven't worked for you, then you are even more likely to want to try one of these kits. Again, you may be able to relate the results to CM and BBT observations and finally make some sense of them. Or you may end up deciding just to use the kits and "go for it," as one of our readers wrote she and her husband did (with success in that case).

Once again, we don't recommend using the kits as your sole means of determining ovulation time because, as discussed earlier, these tests are *not* fail-safe. Still, it is somewhat easier to use them when trying for a boy. And if you decide you must rely on them *solely,* here's what we suggest you do:

First of all, *test twice a day without fail.* Read the instructions in your test kit very carefully so you know how to interpret the results properly. When the test strip changes color, indicating that the LH surge has begun, you want to be sure that it really has just recently begun (within a few hours). Again, test between 11 A.M. and 3 P.M. and then between 5 P.M. and 10 P.M. Space the two tests apart by at least eight hours. Noon and 10 P.M. would be ideal testing times.

Let's say that when you test during the midday period, you get negative results—that is, there is no indication of the surge. Then you test again, in the evening, let's say at 8 P.M. This time there is a positive result—the LH surge is under way. You can assume that the surge actually occurred at least five or six hours earlier (since it is not detected in the urine until several hours after it actually occurs in the blood) and that, in most instances, ovulation will occur about thirty hours after the surge hits the blood. Thus, ovulation in this case is likely to occur about twenty-four hours after this test—in other words, at about 8 P.M. the following evening.

Our recommendation in a case like this is for the couple to have unprotected intercourse sometime between the following morning—about twelve hours before the suspected time of ovulation—and early evening. If you can split the difference and have intercourse around midday, so much the better.

Remember, you must test twice a day if you are going to use this method in your actual attempt to conceive a boy. And you must wait twelve hours after you first detect the LH surge before having intercourse. Intercourse immediately after detecting the surge could increase the possibility of conceiving a girl.

Timing Summary

Here is a summary of timing techniques for the boy. In terms of BBT, try for the boy between the morning on which you observe what you believe to be the last low temperature and the morning on which you note the marked rise in temperature. Achieve finer tuning and greater accuracy by using your analysis of the CM. Try for the boy only on those days when you are experiencing peak symptoms. If you have more than one consecutive day of peak symptoms, the last day is the most likely day of ovulation. Use common sense, however. A few women with hormonal disturbances, which may be of only a transitory nature, sometimes exhibit clear, fertile-appearing mucus at odd times during the cycle. When things seem to be out of kilter with your normal experience, be particularly cautious and seek additional information through the use of BBT, *Mittelschmerz,* and the ovulation-prediction kits.

Your best chances of conceiving a boy arise when intercourse is timed as close as possible to the shift from peak mucus back to thicker, cloudier mucus. Ovulation actually may occur sometime before rather than at the time the mucus shifts back, so we don't recommend having intercourse for the boy after this shift is noted. You still may have a greater chance of conceiving a boy for up to twelve hours *after* ovulation, but data on this remains ambiguous. It is always best to try to ensure that intercourse takes place *before* ovulation occurs. Once the mucus changes, it will be more difficult for the male-producing sperm to get to the egg first.

The Rules of Abstinence

Factors other than timing can enhance chances for conceiving male offspring. In the past, Dr. Shettles recommended sexual abstinence from the beginning of the cycle until the day on which sex selection is attempted. The reasoning behind this was that higher sperm counts have been associated with a higher incidence of male offspring, as discussed earlier in this book. By abstaining from sex until the target date, near the middle of the cycle, Dr. Shettles reasoned, the husband's sperm count would be maximal.

A number of couples have complained that this is too long a period to abstain. We have come to agree, particularly since that long an abstinence might be counterproductive. After so long an abstinence, spontaneous nocturnal emissions may occur, possibly just before the time when sex selection will be tried. That would have the effect of *lowering* sperm count at just the wrong time.

Dr. Shettles now advises that it is all right to have intercourse during the first days of the cycle, but, even here, it is *absolutely essential that condoms be used.* If you don't use condoms and you are off at all in your timing, you could conceive a girl. When you believe you are within four days of your time of ovulation, *abstain entirely* from any sexual activity that results in male ejaculation. It is important to increase sperm count as much as possible—a factor that favors male conceptions. (We've heard from a few men who say they can't abstain that long without having a spontaneous ejaculation. If this is a problem, then abstain beginning three days before suspected ovulation time.) Have intercourse without a condom only on the suspected day of ovulation.

We recommend having intercourse only once on the day of suspected ovulation. After having this one unprotected intercourse, you *must resume using condoms* for the next several days, to be entirely safe. It is especially important to use condoms for the first three days after ovulation has taken place.

"Keep It Cool" and Other Good Advice

Some people thought Dr. Shettles a tad daft several years ago when he began telling men who want sons to avoid jockey shorts, jockstraps, and other tight-fitting clothing, all of which can have the effect, he warned, of raising the temperature within the testes to the point where sperm count is reduced. Some men wore such tight-fitting clothing, he found, that they actually rendered themselves temporarily infertile. Heat kills both types of sperm, but it kills the smaller, less protected, male-producing type first. For men who want sons, Dr. Shettles' recommendation is "Keep it cool."

Dr. Shettles has made a considerable study of the relationship among scrotal temperature, spermatogenesis, and the sex of offspring. He found that since the time of Hippocrates, hyperthermia (excessive heat) has been recognized as being injurious to reproductive functioning in the male. Reduced conception rates have been related to a number of heat sources, including tight-fitting jockey shorts, insulated underwear, overheated workplaces, and so on. Taking too many Turkish baths or soaking for too long in a hot tub can also lower sperm counts.

Studies are few on the relationship between heat and sex ratios, but those that exist tend to support Dr. Shettles' argument that resulting lower sperm counts yield an excess of girls. Skin divers who spend a great deal of time in tight-fitting, heat-retaining rubber wet suits, for example, have been shown to father far more daughters than sons.

Some men may be unaware that their testes are, in fact, being exposed to abnormally high temperatures. It doesn't take too much of a temperature increase to affect sperm production. Men who work in furnace rooms or in front of large industrial ovens are at risk; so are men who sit all day in heated vehicles, such as taxis or trucks. These are the kinds of men, Katherine Bouton reports in an article ("New Light on Male Infertility," *This World*, July 11, 1982), whom infertility doctors are seeing more and more of. Men who work in bakeries, pizza parlors, and the like are also being seen in increasing numbers.

Dr. Shettles notes that men who jog in plastic suits (designed to increase weight loss) are at real risk. Even frequent saunas can lower sperm counts. Fortunately, once the heat is off, the sperm count gradually increases again. Men who have such low counts that they are infertile regain their fertility within a few months of avoiding whatever heat source caused their problem in the first place. Some fertility experts have recommended placing the testes on ice or immersing them in cold water to help speed things along. That isn't one of Dr. Shettles' recommendations (unless your doctor has suggested it and is supervising the process), but he does advise avoidance of anything that causes excessive heat to be generated in the scrotal area.

One woman who wanted a son wrote us that she had hidden her husband's tight-fitting jockey shorts. Not a bad idea. Try boxer shorts.

Be aware, too, that *any* form of stress, not just heat, can lower sperm counts. Protracted illnesses, even a bout of the flu, can do this. Exposure to toxic chemicals reduces sperm counts, and, as with heat, there are studies showing that exposure to such chemicals can result in an excess number of female conceptions. Radiation lowers sperm counts; so do many drugs, both prescription and nonprescription. It has been established that prolonged use of marijuana can reduce male sex hormones and lower sperm counts.

Actually, there is one drug—a mild one—that may actually help you conceive a boy: *coffee.* This is purely optional, but drinking a couple of cups of strong caffeinated coffee fifteen to thirty minutes before having intercourse, on the try for the boy, may impart some extra speed to the male-producing sperm. A number of researchers have shown that caffeine can have a stimulating effect on sperm. It gives both types a boost but, according to Dr. Shettles, gives the male-producing type the greater boost. It won't do any good for the wife to drink the coffee; the husband has to do it. And we *don't* recommend long-term heavy use of coffee, which *might* actually have the opposite effect—a lowering of sperm count.

Since we first proposed this coffee boost in the previous edition of our book, we've heard from numerous couples who say they can't stand coffee and wonder if strong black tea or some other source of caffeine might not do as well. These other caffeine sources are fine. Again, we don't recommend *any* high-caffeine food or drink for long-term use. Women—see the next chapter—should avoid caffeine.

The Alkalinity Factor and Orgasm

The chemical environment of the female reproductive tract is very important in sex selection. Most women go through cyclical changes, with that environment becoming more alkaline and more receptive to sperm as ovulation approaches. A few women, however, have highly acidic secretions most of the time, making conception of either sex difficult and male conceptions especially difficult. Some women who have continual "acid stomach" or ulcers may be keenly aware of their overly acidic conditions. Others can assess their status by paying attention to their cervical mucus. If it is always thick, cloudy, not very stretchable, and in poor supply, then the environment is not going to be conducive to male conceptions. If the problem is severe, a doctor's help may be needed to get the system operating properly.

In the past we have recommended baking soda douches to help enhance alkalinity just prior to having intercourse in the actual attempt for the boy. These douches have been used safely for decades by thousands of women and recommended by hundreds of doctors for a variety of reasons. In general, however, these douches aren't necessary except in a few cases.

Alkaline douches—or other medical intervention—may be indicated *only* in those women who suffer from frequent vaginal infections (which tend to produce or accompany highly acidic secretions) and some forms of infertility. It is impossible for Dr. Shettles to prescribe for such women through the medium of this book. If you think you are a candidate for the kind of intervention that restores proper acid/alkaline balance to secretions, *consult your doctor and douche only according to his instructions.*

In any case, there is something most of you can do to further enhance the alkalinity and penetrability of the cervical secretions—something entirely natural. This "something" is female orgasm. The woman trying to conceive a boy should also try to experience orgasm during the critical intercourse. And if at all possible, she should try to have it at the same time as or, better yet, just *before* the male orgasm. Female orgasm usually increases the quantity and flow of the natural alkaline secretions that arise close to the time of ovulation. The orgasmic contractions of the female, moreover, help rapidly transport sperm into the cervix, where the secretions tend to be even more favorable for the male-producing sperm. Multiple female orgasms that precede male climax are better yet, but don't knock yourself out trying to achieve them. Female orgasm is helpful but by no means crucial.

(See "Questions and Answers" for more information on douching and a method for improving quality of cervical mucus.)

Position and Penetration

Two more factors favor male conceptions. The man should try for *deep* penetration at the time of his climax. This will help deposit the sperm closest to the cervix, where secretions are more favorable to the male-producing sperm. In addition, Dr. Shettles recommends vaginal penetration from the rear when trying for the boy. This may sound odd, but, in fact, this position helps ensure that the sperm are deposited near the opening of the cervix rather than in the two spaces adjacent to it, the so-called cul-de-sac and the posterior fornix.

Most couples find the procedures are only a minor inconvenience. Some even write us that they add a little "spice" to their sex lives. And, of course, we have made things easier now by modifying the recommendation that seemed to trouble some couples the most—the one that called for abstinence throughout the first half of the cycle. Four or five days of abstinence prior to ovulation is, as noted above, adequate.

Checklist

Before you try for the boy, be sure that you understand everything related to:

- Determining time of ovulation, using CM, BBT charting, and so on
- Timing of intercourse relative to the time of ovulation
- Abstinence and use of proper birth control
- Importance of sperm count and factors that affect it
- Means of influencing the acidity/alkalinity of the female reproductive tract
- Significance of female orgasm
- Position during intercourse and male penetration

If you *aren't* confident that you've grasped the entire method, don't be impatient. Go back and study the previous chapters until you are sure you are ready. (The next two chapters may answer any further questions you may have.)

We believe that 80 to 85 percent of all couples who correctly determine their time of ovulation and conscientiously follow the instructions for conceiving males will succeed.

Special Problems (Sex-Selection Counseling)

Some couples, for one reason or another, decide that they need help in selecting the sex of their children. Some men are certain that they produce sperm of only one type because of family histories showing a preponderance of children of one sex. Other couples have trouble conceiving at all and would like to combine an effort to overcome infertility with an effort to beget a child of a particular sex. Still others have despaired of finding their time of ovulation and want help with that. Some want to avail themselves of any fail-safe method that might exist or want to be referred to a sex-selection clinic.

Here are some of Dr. Shettles' answers to such inquiries and requests. First of all, there are no centers, other than those that use artificial insemination and PGD, that are devoted to sex selection. If you are having problems with infertility, you should seek a referral from your own physician. If you are having difficulty determining when you ovulate, after having conscientiously tried all of the methods suggested in this book, your own physician may be able to assist you. More and more doctors are becoming interested in sex selection but many still resist it, so you should not expect your doctor to be enthusiastic about helping you in that respect.

Very few men produce sperm of only one type. Even when family histories show a preponderance of children of one sex, there is a good chance that, by following the instructions we have provided, you will be able to conceive a child of the other sex. Only a few doctors in the world would be able or willing to analyze your sperm (using fluorescent dyes and so on) to see whether you do, in fact, produce both types of sperm. Such analyses, moreover, would be expensive. More specialists, however, are now willing and able to analyze sperm, to check their numbers and motility, factors that figure in overall fertility.

Questionnaire

Don't forget to fill out the questionnaire at the end of this book. When you purchase the book you should fill in the first part of the questionnaire, telling us the approximate date you purchased or received the book and your intention to try for a boy. Then after your baby has been born, fill out the rest of the questionnaire, letting us know if you succeeded.

Trying for the Girl—
What to Do

The first thing you may notice is that this chapter is shorter than the preceding one—on the boy. You may conclude from this that it is therefore easier to conceive a girl. Actually, the opposite is true. In many cases it takes a little more patience and effort to conceive a girl. The preceding chapter is longer only because, in it, we reemphasize and further explain some of the procedures for timing ovulation. It is essential that those interested in conceiving a girl read this book in its entirety, *including* the chapter on the boy, before making any sex-selection attempt. The boy chapter, apart from giving you more insights into the relationship between timing of intercourse and ovulation, will alert you to some of the things you *shouldn't* do in the course of trying for the girl.

The method for conceiving girls is a bit trickier than that for the boy because, as Dr. Shettles' work has shown, it is necessary to stop having intercourse two or three days before the day of ovulation. Here you have to be a good prognosticator. The success rate for the girl is somewhat lower than for the boy, but not substantially lower.

Timing for the Girl

Here is the timing strategy for the girl. You will recall that if the egg is in the fallopian tube or about to descend into it, then intercourse at that time (at or near ovulation) will favor the smaller, faster, male-producing sperm. Therefore, for the girl, Dr. Shettles advises couples to have intercourse four, three, or two days *before* the suspected day of ovulation. Then, when the egg finally arrives, it is likely that only the sturdier, longer-lived, female-producing sperm will still be "in waiting" and capable of fertilizing the egg.

The idea is to time intercourse well in advance of ovulation, as far in advance as possible (to begin with) but not so far as to completely rule out the chances of conceiving at all. Those who want to be extremely cautious may want to start at the outer limit—four days in advance of ovulation. Generally, however, Dr. Shettles recommends starting at three days in advance of ovulation because at four days, the chances of conceiving are not great. If three days don't result in pregnancy, you can move to two and a half days and then, finally, to two days. If you have intercourse closer to ovulation than two days, you will place yourself in distinct jeopardy of conceiving a boy.

The first task, as with the boy, is to determine your probable ovulation time. You will find instructions on how to do this in the chapter "First Order of Business: How to Determine the Time of Ovulation," with some additional pointers in the chapter "Trying for the Boy." Here are some things to be particularly mindful of in planning for the girl. During your practice cycles it is imperative, of course, that you use some form of birth control. Do not use the Pill, for reasons explained in the last chapter. You shouldn't use IUDs, either. And though in the past we've indicated it was all right to use contraceptive foams or jellies (with diaphragms), we no longer recommend their use.

It is true that pregnancies that occur as a result of the failures of contraceptive foams and the like tend more often to be girls than boys; this is due, Dr. Shettles believes, to the fact that these chemicals are highly acidic and thus tend to kill the male-producing sperm first. The use of these foams and gels, however, will interfere with your important CM readings and will render use of Tes-Tape worthless. You might say, "So what? If I fail while using foam I'll probably get a girl anyway." Your reasoning would not be sound. In the first place, your chances of conceiving a child of either sex are greatly diminished while using these substances, and, second, you might, if you did manage to conceive at all, still have a boy.

The best form of birth control to use during your practice cycles is the condom. It will keep the vagina free of seminal fluid, which might be mistaken for cervical mucus and thus give you false CM readings. Use condoms and go through enough cycles to make you feel confident that you can predict when ovulation will occur. Then and only then should you make your first attempt.

Concentrate first on mastering the CM method of finding ovulation time. Keep careful charts and notes. Even if you are not particularly successful in pinning down the exact time of ovulation, chances are excellent that you will be able to distinguish between the various gradations of CM. Avoid intercourse, in particular, on your peak CM day, as that is likely the day of ovulation. If your cycle is fairly regular, you will be able to determine when peak symptoms occur. Intercourse should take place at least two days before the peak. Of course, if after several months of trying you still have not conceived, you may decide to risk moving in a bit closer to the peak. And especially if you have two consecutive peak days, eventually you may decide it is necessary to have intercourse on the day prior to the *first* of those peak days. Use caution, patience, and common sense.

You should also learn the BBT method of finding your ovulation time. Use this temperature method in conjunction with the CM method. Place your higher trust in what the CM tells you, unless, for some reason, you find it difficult to interpret the CM but discern fairly clear BBT patterns. Then you may decide to rely more heavily on the BBT. But, almost always, the two methods complement one another. Some may want to consider using an ovulation-prediction kit along with CM and BBT. You'll find instructions on using the kits later in this chapter.

When using the BBT—and this is very important—regard the last low temperature before the sustained rise in temperature as the probable time of ovulation. It may actually occur a little later, but to begin with, err on the side of caution. Schedule your last intercourse for three days prior to that. For example, if you regularly record this last low temperature on day 14 of your cycle, schedule your last intercourse for day 11. If you don't conceive on that schedule, after trying two or three times, you may want to move to day 12.

Two Formulas for Irregular Cycles

If you have particularly irregular cycles and are relying largely on the BBT method, there are two formulas that might help you decide on which day of the cycle you should cease having intercourse. Here are two case histories that will help explain the formulas.

Case 1. Linda's periods are very irregular, and she has never ovulated sooner than day 19. Taking her temperature every morning before she gets out of bed works fine. She always goes from about 97.6 degrees to 98.4 degrees when she ovulates, but one month she ovulated on day 19 and the next on day 25. Her cycle runs anywhere from twenty-eight days to sometimes forty-five. She wonders if there is any hope that she could conceive a girl. She thought perhaps she would have intercourse on day 17 for five or six months and then move on to day 18, 19, and so on, until perhaps she could get pregnant with a girl. She is in no real hurry and would wait four or five years if she knew she could have a girl.

Dr. Shettles suggested that Linda try out two formulas to see which, when applied to her case, would yield the *earliest* day on which she should cease having intercourse. The first formula is simply to look at the records of four to six cycles (preferably more if you have time and your cycle is highly irregular), take the earliest date on which you ovulated during those cycles, and subtract three days. In other words:

earliest ovulation date − 3 days

Using this formula, Linda would subtract three from nineteen and come up with day 16.

The second formula is to take the number of days in your shortest cycle and subtract fourteen (inasmuch as fourteen days often, but now always, elapse after ovulation before the new menstrual period begins);

when you do this, you get the earliest probable ovulation date in *any* month. Then you subtract three days from that to put you at a time that favors female offspring. The formula is:

number of days in shortest cycle − 14 days − 3 days

In Linda's case, then, she would subtract fourteen from twenty-eight (the shortest cycle she has noted), which would give her fourteen. From the fourteen she subtracts three to come up with eleven. Using this formula, then, Linda learns that she should have intercourse up to and *on* day 11 but no later—for her earliest and most cautious attempts to have a girl.

Linda, of course, might only rarely have a cycle as short as twenty-eight days. But since she has no way of knowing in advance, given her great irregularity, and since she is determined to have a girl and has plenty of time to try, Dr. Shettles definitely recommends that she start with a target date of day 11. She and her husband can have frequent intercourse, without a condom, right up to and on day 11. They must have intercourse *on* day 11 and then abstain thereafter, at least until they know they are in a "safe" period. Even then, given the irregularity of Linda's cycles, they are probably best advised to use condoms.

Linda should definitely not *start* with a target date as late as day 17. If after three or four months she fails to conceive at day 11, she can advance to day 12, wait another few months, move on to days 13, 14, 15, and so on, at her own pace. Most women, of course, won't have these extremely irregular cycles and will not have to put forth this much effort. (And, remember, the CM method may give you valuable additional information that will help you clear up some of the ambiguities, although in extremely irregular cycles the CM pattern may be difficult to interpret.) Even the "Lindas," we believe, can be successful—if they can be patient and slowly "creep up" on the objective.

Case 2. Mary is far less irregular than Linda, but after six months of temperature-charting (for one reason or another, she has not wanted to use the CM method), she is convinced that she may ovulate anytime between day 11 and day 15, which is still quite a spread over so short a period of time. The shortest cycle she has recorded is twenty-six days.

The earliest ovulation time she has noted is day 11. Given that information, let us again apply both of the formulas and see which will yield the earliest "cutoff" day for intercourse.

Mary subtracts fourteen from twenty-six and arrives at twelve; she then subtracts three days from that and gets nine. This is the formula that yielded the earliest cutoff day in Linda's case. But let's see what happens when Mary applies the other formula. She subtracts three from her earliest recorded ovulation date—day 11—and gets eight. This formula, then, yields an earlier cutoff date, and so this is the one Mary should use to begin with. If Mary fails to conceive on this schedule, she can gradually move things ahead, timing the last intercourse, for example, on the evening of the eighth day rather than on the morning of that day. If that doesn't work, she can move to the morning of day 9 and so on.

Use of the Ovulation-Prediction Kits

If you decide to use one of the ovulation-prediction kits to help you plan the timing for the girl, first reread what we had to say about these kits in the last two chapters. And use these guidelines:

First, if your cycles are quite regular, you may feel you have no need to use these kits—and you really need not. If your cycles *are* fairly unpredictable, however, and/or you want extra assurance that you are timing things properly, then you may want to consider these kits. (We hope that, by the time this new edition of our book appears, some more reasonably priced kits will be on the market. Already there are differences in price, so it *does* pay to compare products. All of the kits that use monoclonal antibodies are about equally reliable, so far as we can determine.)

It's best not to start using these kits—given their cost—until you have some idea of when you ovulate. Then start four or five days ahead of that date and see if you are right. Remember that when you pick up the LH surge, you are, in most cases, within a day or a day and a half of ovulating. Many couples trying for a girl have written to ask us if it is all right to continue to have unprotected sex right up until the test indicates the LH surge is under way. Our answer is *no*—because, as we've just stated, having sex right up to the surge could still put you too close to the "boy zone" for comfort. This is particularly true if you test only once a day, since by the time the LH surge is detected by that single test, it may actually have started many hours before—possibly a full day earlier.

It's okay to test only once a day to try to get additional information during your practice cycles, but if you are really going to rely in any substantial way on these ovulation-prediction tests for your actual attempt to conceive a girl, then we *strongly recommend that you use them twice a day.* And just as we suggested in the last chapter, we advise that you test between 11 A.M. and 3 P.M. and again between 5 P.M. and 10 P.M. Space the two tests at least eight hours apart for best results.

In the majority of cases it appears that ovulation occurs about thirty hours after the LH surge takes place in the blood. The LH doesn't show up in the urine for another four to six hours, typically, so even if you detect the surge when it first enters the urine, you're still a little "after the fact" and only about twenty-four hours from ovulation in many cases. Thus those trying for girls will want to *stop* having unprotected sex at least a day *before* detecting the LH surge (while testing twice a day).

When it comes to trying for the girl, these prediction tests can best be used during practice cycles. Using BBT and CM, choose a safe cutoff date and then use the kit for the next few days to see how accurate your choice was. For the most part you'll have to continue relying on BBT and CM, but these kits may prove useful to some of you. Be advised, however, that if you are experiencing "peak" CM symptoms, you would be at real risk of conceiving a boy no matter what your ovulation-prediction kit might tell you to the contrary.

Recent studies show good correlation among the prediction kits, BBT, and CM—when all are used properly. Still, there are many individual variations—so get as many "opinions" as possible. When in doubt, use our tried-and-true data related to CM and BBT.

Incidentally, one study found that the LH surge is most likely to occur in the morning hours in the spring and summer and in the evening during fall and winter. That needs confirmation, but it is intriguing. If you follow our twice-a-day testing schedule, you'll be pretty well covered in any case.

Essentials of Timing, Abstinence, and Birth Control

Here are the essentials covered so far. Use condoms throughout your practice cycles. During those cycles in which you make actual attempts to conceive the girl, dispense with the condoms and have intercourse daily, if possible, from the time bleeding ceases right up to and on the cutoff day—four, three, or two days prior to the suspected time of ovulation. After the cutoff date, abstain from intercourse entirely, even with condoms, until a few days after the ovulation has taken place. There is the chance that intercourse, even with a condom, during the ovulatory phase will further stimulate female secretions that might favor male-producing sperm. When you resume intercourse two or three days after ovulation, use condoms again to be on the safe side.

Why have intercourse frequently from the time bleeding ceases up through the cutoff date? Dr. Shettles recommends this so that the sperm count will be naturally diminished, a factor that he has found may favor female conception. *Don't* attempt any artificial means of reducing sperm count. This could be dangerous and counterproductive. (The sperm-diminishing factors discussed in the previous chapter, such as excessive heat and radiation, often result in male infertility.)

To Douche or Not to Douche?

In previous editions of this book a douche was recommended to impart a more acidic character to the secretions. Further study has indicated that these douches are generally unnecessary. If you have fertility problems, infections, or other difficulties that you think may be helped by douching, consult your physician and follow his or her instructions. No adverse effects were ever reported with the douches that were used in the past (thousands of women have used them over many years for a variety of reasons), but since equally good results can be obtained without them and since we cannot know the precise condition of every woman using them, we have discontinued their use in the Shettles method and now advise women to douche only if so advised by their doctors. (See "Questions and Answers" for more information on douching.)

Sexual Position, Penetration, and Orgasm

Dr. Shettles recommends that the "missionary position" be assumed during intercourse for the girl. This is the face-to-face position, with the male on top, a position that makes it less likely that the sperm will be deposited directly near the opening of the cervix, where the secretions are most alkaline and would favor the male-producing sperm. Shallow penetration by the male at the time of his orgasm will further help ensure that the sperm passes through the vaginal canal, where secretions are naturally more acidic and therefore, in a relative sense, favor the female-producing sperm.

If the woman can avoid orgasm while trying for the female, that can help, too, Dr. Shettles asserts. Female orgasm increases the flow of alkaline secretions, aiding the male-producing sperm. And the contractions that accompany the female orgasm help propel the sperm up and into the cervix, also giving greater assistance to the male sperm.

Women: Avoid Caffeine

When it comes to caffeine, what's good for the gander is *not* good for the goose. Studies released in 1995 reported that caffeine significantly impedes overall female fertility. A Johns Hopkins University study found that women who consume 300 milligrams of caffeine per day (about the amount you get in three cups of regular coffee) are 26 percent less likely to conceive than are women who consume no caffeine. Even drinking one or two cups of coffee a day reduces the chances of conception by about 10 percent.

Other recent studies have confirmed this and have further indicated that caffeine consumption can interrupt the menstrual cycle and lead to early pregnancy loss in some women. (Smoking and alcohol consumption also reduce fertility in both men and women.) We don't recommend sustained high caffeine consumption for men, either. (See preceding chapter for caffeine guidelines for men.)

Checklist

Before you begin, make sure that you understand everything related to:

- Determining your time of ovulation (using BBT, CM, etc.)
- Timing of intercourse relative to the time of ovulation
- Rules concerning abstinence and use of condoms
- The sperm count factor and frequency of intercourse prior to the "cutoff"
- Sexual positioning and male penetration
- Withholding of female orgasm

If you have any doubts, reread this chapter and the two preceding ones. You may find answers to other questions in the next chapter.

We believe that 75 to 80 percent of all couples who correctly determine the time of ovulation and follow our instructions carefully will succeed in conceiving girls.

Special Problems (Sex-Selection Counseling)

See the end of the previous chapter for information on this topic (under the same subheading).

Questionnaire

Don't forget to fill out the questionnaire at the end of this book. When you purchase the book, you should fill in the first part of the questionnaire, telling us the approximate date you purchased or received the book and your intention to try for a girl. Then after your baby has been born, fill out the rest of the questionnaire, letting us know if you succeeded.

Questions and Answers

Here are some of the thousands of questions we have received over the years. Many of them have already been answered at some point in this book. Some of this information, however, bears repeating.

Q: After three boys in a row, we desperately want to have a girl for our fourth and final child. We read about your method in the newspaper. Can you guarantee it will work for us?

A: No. This method is not "fail-safe." There are, in fact, *no* fail-safe methods of preconception sex selection. You should not plan to have a fourth child unless you are certain that you are prepared to want and love it, no matter what sex it turns out to be. Also, we would caution you against trying to use *any* sex-selection method based only on your reading of a newspaper report.

Q: My husband comes from a family that has a long history of producing mostly female offspring. We've already had two daughters ourselves and, of course, would like to have a son. We're afraid to try, though, without first having a sperm analysis. Our doctor is not helpful. What do you advise?

A: The kind of sperm analysis you refer to can be expensive. The fact is, however, that it is extremely rare for a man to produce sperm of nothing but one type. We have heard from a number of people who have histories like your husband's but who have, nonetheless, succeeded in conceiving a child of the sex they desired by following

Dr. Shettles' instructions—the same instructions that are contained in this book.

Q: We don't understand why we failed. We followed your instructions for the boy to the letter, including the coffee, multiple female orgasm, and so on. The only thing we may have been off a bit on was the timing, but we were sure that by doing all the other things we would succeed. My charts are enclosed. Can you tell us why we failed?

A: It appears evident to us, from your charts, that you had intercourse too early in the cycle. We have heard from a number of people such as yourself, who say they have done everything correctly except, they acknowledge, the timing. We have always stated that timing is not just *one* of the elements you need to adhere to; it is *the* element that is most important. You can dispense with everything else and still have a good chance of succeeding if your timing is just right. If it isn't, your chances of success are slim, no matter how closely you follow all the other recommendations.

Q: We've been trying for the girl, and since you recommend frequent intercourse for the girl, up to two days before ovulation, that means an awful lot of douching—before each intercourse. Is it really necessary?

A: We no longer recommend douching at all. Accumulated data now make it evident that equally good results can be obtained without douching, whether trying for the boy or the girl. Consult your physician before beginning any regular regimen of douching.

Q: I have very acidic secretions and would like to use an alkaline douche to help ensure that we conceive a boy. What do you recommend?

A: Use douches *only* with your personal physician's prior approval. Discuss with him or her the possibility of using a baking soda douche just before intercourse. Such douches typically use two tablespoons of baking soda thoroughly mixed in about a quart of clean, warm—not hot—water. The douche bag should be of the gravity type so that it can be suspended from the shower rail and flow gently into the vagina while you sit or recline in the tub. Move the nozzle carefully and do not insert it too deeply. Do *not* use the bulb-type prepared douche. It can exert too much intrauterine and tubal pressure.

We used to routinely recommend alkaline douches for the boy method. We no longer do so, although we feel they can be quite useful in many women. They provide for a more hospitable environment for male-producing sperm. And, in fact, many doctors recommend alkaline douches for couples having trouble getting pregnant at all because the alkaline environment helps sperm in general to get through the secretions to the egg. But since the male-producing sperm are faster, the alkalinity helps them *more*.

Even though women have been using alkaline douches for decades, there have been a few reports that douching may be associated with pelvic inflammatory disease in some women. The theory is that douching, and especially frequent douching, may result in an imbalance in the vaginal environment that increases vulnerability to infection under some circumstances. This association has not been proved, but we believe that a woman should *not* douche without the prior approval of her personal physician.

Q: We've heard about a sex-selection method based on diet. What do you think of it?

A: We think it is interesting and shows some promise. Unfortunately, the proposed diets pose some potential health risks and so we can't recommend using the diet method, even as an adjunct to the Shettles method at this time. We hope the diet method might be refined over the years, to remove any health peril. For more information on this method, see the chapter "Are There Other Methods? Are They Conflicting?"

Q: We've heard that your method is based entirely on the results of artificial insemination and that they might not apply to normal reproduction. Is this true?

A: No. The artificial insemination data is supportive of the Shettles method in that babies conceived via artificial insemination, where care is taken to time the insemination as close as possible to ovulation, are far more likely to be boys than girls. And it is true that, to some extent, the Shettles method mimics the conditions of artificial insemination. But the Shettles success rate is based on inseminations via ordinary sexual intercourse.

Q: I have very scanty cervical secretions. I've heard that a cough medicine can improve the situation. Is this true and, if so, what is it? I would like to use it to try to help in our effort to conceive a boy.

A: Guaifenesin is an expectorant that works by increasing fluid in the respiratory tract. It turns out that this same substance improves the quality of cervical fluids considerably. It has helped overcome infertility in a significant proportion of couples studied. Thus it might be helpful, as you have suggested, in the quest for male offspring. Dr. Shettles has used it in some patients, advising them to take one teaspoonful of over-the-counter *plain* Robitussin (cough medicine that contains guaifenesin as its active ingredient) three times a day for three or four days before trying for the boy. It should not be taken longer than that or in greater quantities than that, and you should consult with your physician before using it. And do *not* use the form of Robitussin that contains codeine. Check labels carefully. Use the form *without* codeine. A study in *Fertility and Sterility* found guaifenesin to be very helpful in making cervical fluids more hospitable to sperm, a condition that favors the male-producing sperm and fertility in general.

Q: I am writing to ask why you don't put more emphasis on the Billings method of charting the cycle. I have used this for years as a method of birth control and find that it works far better than the temperature charting. Too many little things can affect the temperature. But with the Billings method I'm always sure of my fertile days and especially my ovulation day. Why not give more information on this excellent method?

A: We have done exactly that in this book. In the past we've relied primarily on the temperature (BBT) method of charting cycles. Now we think, as you do, that the CM (cervical mucus) method (also known as the ovulation method and the Billings method) works better. See the chapter "Trying for the Boy—What to Do" for more details.

Q: *Is it possible to ovulate on different days of the month?*

A: Yes. Whereas some women show a high degree of ovulatory regularity, others do not. Even women who have fairly regular menstrual cycles find that they ovulate on different days, though usually within a narrow range, from cycle to cycle. Do *not* try to rely on calendar calculations to figure out your time of ovulation. You must use the CM and BBT methods and other symptoms of ovulation to do this. See the chapters "First Order of Business," "Trying for the Boy," and "Trying for the Girl" for details.

Q: *Why fool around with BBT and CM since there are now ovulation-prediction kits on the market that can pinpoint ovulation with certainty? Why do you not recommend these?*

A: We discuss the kits and do, in fact, recommend them under certain circumstances. We cannot emphasize too strongly, however, that these tests are *not* surefire and *cannot* "pinpoint ovulation with certainty." When properly used and interpreted, they help predict when ovulation most likely *will* occur. Please read what we have to say about these kits very carefully before using them in any sex-selection effort. The kits are a nice addition, but we continue to believe most couples should rely primarily on the tried-and-true BBT and CM methods.

Q: *Do you still recommend the Tes-Tape?*

A: Yes, but only as an adjunct to the CM and BBT methods of determining the status of the cycle. And you can't use the Tes-Tape during the actual trial for the girl because the seminal fluid will give you false readings. You can use it, however, during your practice cycles when condoms are used. See the chapters "Trying for the Boy" and "Trying for the Girl."

Q: Do all women ovulate on the fourteenth day of their cycles?

A: No, this is an average. Women who have highly regular, twenty-eight-day cycles may find that they ovulate on day 14 during most cycles; other women, however, may ovulate on other days. Do not use averages to try to determine your own ovulation time.

Q: Is it true that most women ovulate fourteen days before the end of their cycles?

A: There may be more truth to this than to the idea that women always ovulate fourteen days from the *beginning* of their cycles. When you examine a very large number of charts, you do find that quite a significant proportion of women *do* ovulate two weeks before the *end* of their cycles. But, again, this is a generality, and you want to rely on *specifics*, not generalities, in planning your own sex-selection strategy.

Q: We want a girl. We understand that we must stop having intercourse three days before my ovulation. When can we resume sexual relations?

A: Actually, you can stop as close as two days before ovulation. As for when you start again, we advise waiting until three or four days after you believe you ovulated. Even though you would be using a condom during the period of suggested abstinence, there is the possibility that sexual activity at that time could upset a conception in progress. In addition, sexual intercourse may result in female orgasm and an increase in alkaline secretions, which could favor male-producing sperm. So: Have intercourse daily, if you like, from the end of bleeding up to two or three days before ovulation during your actual attempt at conceiving a girl. Then have no intercourse whatever until three or four days after ovulation. Even then, to be on the safe side, we recommend you use condoms.

Q: I've recently read that the fertility drug Clomid could be used to help couples have girls. Is this true?

A: It has been noted that women who take Clomid give birth to more girls than boys. Clomid tends to make the cervical secretions less abundant and less alkaline, and this may account for the preponderance of girls. This is a powerful drug that can have adverse side effects, however, and should not be considered for sex-selection purposes. It often results in multiple births. (See the Afterword: "Sex Selection in the Near and Distant Future.")

Q: Does your method work for those who are breast-feeding?

A: No. Breast-feeding alters the natural cycle to the point where the method does not work. Do not attempt to use the method, either for the boy or the girl, until you have ceased breast-feeding for two to three months at a minimum.

Q: We heard somewhere that Drano can be used to figure out if you've got a boy or a girl—after it's already conceived. Is this so?

A: The claim is that if a woman puts a small amount of Drano in a glass and then spits on it, she'll be able to tell whether she's carrying a boy or a girl by observing what happens to the Drano. If it does not change in color, she's carrying a girl, according to the story, but if it turns brown, she is carrying a boy. We've heard reports that at least one New York doctor believes in this test. We don't know if it has any validity at all. It seems highly doubtful. The Drano company itself has denounced the test, and a good many people, including "Dear Abby," have pointed out that it can be a *highly dangerous* test to conduct. One woman spit into the glass and blew some of the Drano crystals into her face and eyes. Another woman urinated into some Drano (urine is supposed to work the same as saliva), and the glass in which the Drano was contained literally exploded. So our advice to you is: Skip the "Drano test." This is potentially dangerous stuff. It should be used only for the purposes for which it was designed and then only according to the precautions on the label.

Q: Should my husband take vitamins to give his boy-producing sperm an extra boost?

A: Vitamins, such as C and E, have been found to promote overall fertility—but they should not be taken in extremely high doses, as these may produce negative effects. Some researchers suggest supplementation with 250 milligrams of vitamin C a day and 400 IUs of E a day—both for men and women. If your doctor recommends prenatal vitamins, by all means take them.

Q: We want to be certain that we get a girl this time. We've been told there are sperm-separation procedures that can guarantee us the sex we want. Where can we go to have this done?

A: Since we have so little confidence in the sperm-separation procedures that are currently available, we cannot recommend any of them. These methods require artificial insemination (which by itself has a fairly high failure rate) and overall, these methods, which are very expensive, time-consuming, and inconvenient, are no more successful than the Shettles method. See the chapter "Are There Other Methods?" and the Afterword for more on sperm-separation techniques. Be advised that we have heard from several couples who tried these sperm-separation procedures and were quite unhappy with what they experienced and with the results.

Q: You may think we're crazy, but we have a method of sex selection that seems to work just fine. We have four children. The boys were both conceived in the winter, and the girls were both conceived in the summer. The first boy and girl just happened that way. We planned the second two. Are we crazy?

A: No, just lucky. Or *probably* just lucky. There have been some studies suggesting that more male conceptions occur in the cool of winter (perhaps when sperm counts are higher) and more girls in the heat of summer. The data here, however, are not very strong. For more information on various notions, some scientific, some not, on the influences of weather and the like, see the chapters "Hundreds of Years of Trial and Error (Mostly Error)," "How Much Scientific Support Is There for the Shettles Method?" and "Are There Other Methods?"

Q: Will using lubricated condoms have any effect on my CM readings?

A: Always use a neutral, water-based lubricant such as K-Y.

Q: Which prescription drugs might interfere with your methods?

A: There are no data on this—but you should *never* use drugs, whether prescription or over-the-counter, while pregnant or attempting to become pregnant without first getting clearance to use them from your physician.

Q: What happens if you have intercourse just after ovulation?

A: If you have intercourse soon after ovulation—within twelve hours—Dr. Shettles still believes the chances of having boys are greater than the chances of having girls. However, we recommend that you time intercourse, for the boy, in the hours just before ovulation. If you are using the CM method of finding ovulation, don't have intercourse after the peak CM symptoms subside and the CM abruptly turns thicker and cloudier again. See the chapters "First Order of Business," "Trying for the Boy," and "Trying for the Girl" for instructions.

Q: I've read that women who take or have taken LSD are far more likely to give birth to girls than to boys. Could this be true?

A: We don't know. One study indicated that this was true, but we have never seen any follow-up done, either to refute it or support it. We do know that *men* who use certain drugs are more likely to father girls. Anything that creates physical or even emotional stress (which, in turn, translates into physical stress) can reduce sperm counts, leaving more of the hardier, female-producing sperm intact while decimating the male-producing variety.

Q: *Can the use of marijuana make a man infertile?*

A: Abuse of any of a number of drugs, including alcohol, can result in at least temporary male infertility. Marijuana has been shown to diminish male hormones and thus reduce sperm counts, factors that might favor female conceptions or result in infertility. Normal fertility can be regained, in most cases, by abstaining from drugs for a period of months.

Q: *Does age have any effect on the sex of the child?*

A: Since sperm count tends to decline with the age of the male, it is possible that older men father more daughters than sons. And since the secretions of the woman tend to become less copious and alkaline with advancing age, it is also possible that these factors contribute to more female offspring later in life. Some studies seem to confirm this. One study showed that women of about fifteen, twenty, thirty, and forty years of age had offspring with sex ratios of 130, 120, 112, and 91 males, respectively, for every 100 females. By following the recommendations in this book, however, we believe that older couples can still select the sex of their children with a high degree of success.

Q: *We have three girls and are willing to have an abortion if we conceive another girl. Can you tell us about tests to determine which sex has been conceived?*

A: We are adamantly opposed to the use of abortion for postconception sex selection. So are most doctors. Amniocentesis can determine the sex of the child, but this cannot be performed until well into pregnancy, when the abortion risks are much greater. The same is true of ultrasound. Some bioethicists have recently pointed out that aborting a child simply because it is of the "wrong" sex can have a very negative psychological impact on the other children in the family, particularly if they are of the same "unwanted" sex.

Q: *As a man, I must object to your statement that the male is solely responsible for the sex of the child. He contributes the male- and female-producing sperm, but isn't it also true that conditions within the woman can favor one or the other type of sperm?*

A: The woman, in that respect, does share in the responsibility. Women who have secretions that are highly acidic much of the time put their husbands' male-producing sperm at a definite disadvantage and thus help "select" female offspring. There are other ways in which women may play the decisive role, too. One woman, who had three daughters and wanted a son, wrote us that she had been told that the best time to conceive was exactly five days after her bleeding ceased. The woman timed intercourse for that date each time she tried to become pregnant—and each time she conceived girls. The reason is not hard to guess. This timing schedule placed intercourse well before her probable ovulation date. The misinformation she was using helped "select" girls.

In another case, where odd factors were at work, it was the man who was responsible for an abundance of girls. He was a truck driver who operated on a regular schedule—one that allowed him to return home only at certain times and then for very brief periods. It turned out that, on this schedule, he never had intercourse with his wife later than the twelfth day of her cycle. His wife ovulated on the fourteenth day. This couple had three girls.

Q: *I've heard about a technique that enables doctors to implant a single sperm into an egg. Couldn't this be used to guarantee the sex of the child?*

A: Yes, it could. This technique, which was reported on in 1993, is being proposed for the treatment of some forms of male infertility. It is still experimental, and expensive. It has not been approved for sex-selection purposes anywhere in the world, but it seems likely it will be used for that purpose eventually. (See the Afterword for more details.)

Q: How long do the sperm and egg live?

A: Sperm rarely live longer than four days. Most live no more than three days. It used to be thought that the unfertilized egg lives only about twelve hours. It now appears that it lives at least twenty-four hours and in some cases longer. Fertilization more than twelve hours after ovulation, however, is not highly likely, in Dr. Shettles' opinion.

Q: You say that you are more likely to have girls if you try two days before ovulation. What about one day?

A: Perhaps if we explain it in terms of hours, that will make it clearer. Your chances of conceiving boys are greater if you time intercourse to occur anytime within the twenty-four hours that precede ovulation. Obviously, though, your chances will be better at twelve hours before ovulation than at twenty-four hours before. Between twenty-four and ninety-six hours before ovulation, your chances of having girls are greater.

Q: Do you have any idea how long it will take us to get pregnant if we never have intercourse closer to ovulation than two or three days? (We want a girl.)

A: It may, indeed, take longer to conceive a girl than a boy. Dr. Shettles reported that one group of twenty-two couples who wanted female offspring took up to six months to conceive by consistently timing intercourse two or three days before ovulation. Of the twenty-two children who ultimately resulted, nineteen were girls. (The success rate in this trial was higher than normally reported, perhaps in part because the couples received medical supervision in timing ovulation.)

Q: I would like to use Dr. Shettles' sex-selection techniques, but, as a Catholic, I am wondering if they run contrary to church doctrine?

A: The Catholic Church does not object to the procedures. See the chapter "Is it Moral? Should We Do It?"

Q: I'm currently on the Pill. How long should I wait after going off it before trying to choose the sex of my next baby?

A: Most doctors say three to six months. Dr. Shettles suggests six months. You need to let your menstrual cycle become regular again before trying to become pregnant; studies have shown that women who become pregnant soon after discontinuing the Pill are more likely to suffer miscarriages and other complications of pregnancy.

Q: I have a tipped uterus. Should I follow a different method from most women?

A: We've heard from a number of women with this problem. And, in fact, women with tipped uteruses *should* follow a modified procedure. The condition in which the uterus is tipped toward the spine instead of toward the stomach (technically known as uterine retrodisplacement) used to be corrected by surgery quite commonly. The doctor opened the pelvic cavity and shortened the ligaments that support the uterus, pulling the organ back into a more normal position. Today these operations are rarely performed; a tipped uterus is of little or no clinical significance, and a woman with a tipped uterus is often unaware that she has one. The condition is rare, but if your doctor has told you that you have a tipped uterus, you can alter the recommended procedures as follows: for the girl, follow the method without any changes. For the boy, follow the method as described, but then, as soon as intercourse is complete, lie on your stomach with a pillow under your upper thighs. Remain quietly in this position for about fifteen minutes. This will help ensure that the sperm immediately flow toward the cervix rather than into the posterior fornix, the space behind the opening of the cervix.

Q: My wife claims to have read that the type of shorts a man wears can affect his fertility. If she weren't a nurse I'd say she was crazy. Is there any truth in this idea, and, if so, could it have something to do with the sex of a man's children?

A: Your wife's report is accurate. Dr. John Rock, one of the developers of the birth-control pill, states:

> *Any clothing that prevents maintenance of an intrascrotal temperature that is at least one degree centigrade below [normal] body temperature will significantly lower sperm output. Daily wear of a well-fitting, closely knit jockstrap results in infertility after four weeks. . . . Normal output gradually is resumed after another three weeks without such interference. Enclosing the scrotum in ice for one-half hour daily may increase sperm output in perhaps 10 percent of moderately oligospermic [low-sperm-count] men and result in a long-awaited pregnancy.*

Don't try the ice, however, without a doctor's orders. Just get rid of the tight-fitting shorts or other clothing that lowers sperm count, a factor that Dr. Shettles has shown favors the conception of girls. (But don't use tight-fitting garments to try to conceive girls, either. For more details, see chapters "Trying for the Boy" and "Trying for the Girl.")

Q: Can the sort of occupation or environment that a person works in have any effect on the sex of offspring?

A: Anything that lowers sperm count, including heat, toxic chemicals, certain drugs, acute psychological stress, and so on, may make it more difficult to conceive a child of either sex and particularly a child of the male sex. The smaller, more vulnerable, male-producing sperm are the first to succumb to such stresses.

Q: Isn't sex selection going to result in a world top-heavy with males?

A: The best studies indicate that this commonly voiced fear is unfounded. See the chapter "Is It Moral? Should We Do It?"

Q: My husband hates coffee—is there something else he can drink when we try for the boy?

A: The coffee is strictly optional to begin with. Other caffeinated drinks, such as regular tea, will work just as well.

Q: I've heard that there is a greater chance of miscarriage if an "old" egg is fertilized. Is this true, and does this present a hazard in sex selection?

A: What is meant by "old egg" here is an egg that is not quickly fertilized. The data related to this are skimpy and controversial. And this hazard has never been demonstrated in humans. But, in any case, the Shettles method could be expected to *reduce* rather than increase this risk, if in fact there is any risk, since it emphasizes having capacitated (fertilization-ready) sperm in the tubes either before or at the same time that the egg arrives. It utilizes a "fresh" rather than an "old" egg. "Mother Nature," it should be pointed out, doesn't take this precaution, allowing, as she does, for intercourse to occur at any time. So, again, if there really is any risk, the Shettles method turns out to be safer than the "natural method."

Q: I have only one ovary. Will I still ovulate every month, or only once every two months?

A: Most likely you will still ovulate each month.

Q: I'm confused. Do I use the Whelan method or the Shettles method? Which has the most scientific backing and makes most sense?

A: See our chapter "Are There Other Methods?"

Q: Help! So far as I can determine, I ovulate during my bleeding period. Is this possible—and what do I do if I'm trying for a boy?

A: It is possible—in *rare* cases. Occasionally, women with very short cycles ovulate near the end of their bleeding period, but, again, this is quite rare. Ovulation-prediction kits may be useful in such cases, whether trying for the boy or the girl.

Q: Does oral sex interfere with acidity or alkalinity of vaginal secretions? Should this practice be abstained from in attempting the Shettles method?

A: Just to be on the safe side, abstain from this practice during those periods when you are attempting sex preselection. It's possible that an infection could be transmitted in this way and adversely alter the vaginal milieu.

Q: Do antihistamines have any effect on sex selection?

A: None that we know of; consult your physician, however, about *all* drug use prior to conception/pregnancy—and *during* pregnancy.

Q: Is it okay to use K-Y jelly as a lubricant? Does this adversely affect your method?

A: It might. We recommend avoiding lubricants, if at all possible, during actual attempts at sex preselection. These lubricants can be used in practice cycles. If you must use a lubricant, however, K-Y is probably your best choice.

Q: Does taking vitamin C make my secretions more acidic?

A: We have no data on this. Small doses are unlikely to have any appreciable effect. When you start taking very large doses, however (500 milligrams daily and up), it is possible this could have some effect. If you have been using "megadoses" of vitamin C, we suggest you cut back during sex-preselection efforts.

Q: I often suffer from vaginal yeast infections. Do these make my system more acidic, and, if so, do they lower my chances of conceiving a son?

A: Vaginal infections do, generally, make secretions more acidic. Consult your physician and get these infections cleared up before attempting sex preselection.

Q: I just don't seem to produce enough CM to be able to test it with tissue alone. Is it all right to "reach in" and test with a finger?

A: Yes, it is all right to "probe" with fingers. Follow the instructions we provided in the chapter "First Order of Business" under our discussion of the CM method of determining ovulation time.

Q: When you talk about various forms of "stress," such as heat, chemicals, and so on, lowering sperm counts, do you also mean just getting stressed out psychologically, emotionally? Can that lower sperm count, too?

A: No studies that we know of have been done specifically on this issue, but, given what we know about the effects of stress on many other bodily functions, it seems likely that emotional/psychological stress—especially when persistent and intense—could be expected to lower sperm counts.

Afterword

Sex Selection in the Near and Distant Future

Many people write us, asking, "What's next in sex selection? Will there be a 100-percent effective method soon?"

Sex selection in the near future, say the next ten years, will probably be much as it is now. Better methods of separating the two types of sperm may enable doctors to improve on current techniques somewhat—but only with methods that use artificial insemination.

We reported in this book on sperm-separation procedures that provide for a high concentration of male-producing sperm but not female-producing sperm. About 80 percent of those who conceive as a result of being artificially inseminated with the male-sperm concentrate have sons, it is claimed. Thus the success rate for this method is no higher than that reported by Dr. Shettles and others who have used this method, a method that does not require artificial insemination.

It should be remembered that artificial insemination, even without the extra step of concentrating the male-producing sperm, yields significantly more boys than girls. This is so because artificial insemination is usually made to coincide as closely as possible with the time of ovulation. Dr. Shettles has found that when he combines artificial insemination with various aspects of his method, he can boost his success rate, for the boy, from the 80 to 85 percent range to 90 percent. He uses a "split ejaculate" (the first part of the seminal emission, which contains a higher concentration of sperm) and mixes this with highly alkaline endocervical secretions; this mixture is then expressed directly into the cervical opening at the laboratory-confirmed time of ovulation.

In a trial of this method, nineteen of twenty-one women thus inseminated conceived boys *on the first try,* a 90 percent success rate. This is very significant because many of the practitioners of artificial insemination find that a large number of women fail to become pregnant, via artificial insemination, after two, three, or even more attempts. But even given these results, Dr. Shettles still feels that artificial insemination is seldom worth the time, cost, and inconvenience when being used for sex selection. The results of the standard method are almost as good.

A couple of researchers have recently reported sperm-separation techniques that provide concentrates of *female*-producing sperm, though these remain experimental. Dr. B. C. Bhattacharya uses a separation technique that exploits the tiny differences in the surface electrical charge of the two types of sperm. In chambers that contain oppositely charged electrical poles, the male-producing sperm (which reportedly have a slightly negative charge) gravitate toward the positively charged pole (attraction of opposites), and the positively charged female-producing sperm are attracted to the negatively charged pole. The method is said to result in concentrations of the two types of sperm with 75 to 80 percent purity.

Another experimental method exploits a genetic (antigen) marker that is more fully expressed on the surface of boy-producing sperm than on girl-producing sperm. This surface antigen can be made to bind with cellulose beads, effectively separating the two types of sperm. Since the female-producing sperm do not bind to the beads, they pass freely through the apparatus in which the separation occurs. This technique is said to yield concentrates of 92 percent purity.

The actual safety and efficacy of these experimental methods, however, will have to await further confirmation and testing, probably in laboratory animals, long before they will ever find clinical application in humans. Do not look for separation techniques such as these to be made available to the public for many years. Eventually, however, it seems evident that satisfactory methods will be perfected, even if these don't work out. Of course, even then, the "fail-safe" or near "fail-safe" method will require artificial insemination.

The technology is at hand today to fertilize human eggs by "microinjecting" a *single* sperm cell under the zona pellucida of the egg. A paper appeared in the medical literature on this (as purely an experimental procedure), prompting Dr. Shettles to note that there are additional techniques by which sperm cells can be immobilized and their X and Y chromosomes revealed without harming them. The combinations of these technologies, he pointed out, could result in "a very exact method for sex preselection" in the not too distant future. It is likely, however, that such a procedure would be reserved for those with sex-linked disorders, since this approach requires not only an entirely new mode of artificial insemination but also test-tube conception and reintroduction of the fertilized egg into the womb, technologies that are costly and are practiced in only a few centers.

There was further progress with this technology when a group of Belgian researchers led by André Van Steirteghem, at the Center for Reproductive Medicine, University Hospital in Brussels, announced success in using the single-sperm techniques to overcome some forms of male infertility. Some couples successfully using the technique had previously tried for years to achieve pregnancy without success. The technique has now begun to be used in this country. Clinics that were reported to be active in the work include the Genetics and IVF Institute in Fairfax, Virginia, Department of Obstetrics at Mount Sinai Medical Center in New York City, and others. But, again, the applications at this time are for infertility, not sex preselection. Researchers are increasingly excited about the procedure because it has the potential to do for male infertility what the various *in vitro* techniques have done for female infertility.

Many scientists expressed surprise that the single-sperm technique works. There are enzymes around the sperm that many thought would have to be removed before a single sperm, directly injected into an egg, would be able to do its work. Dr. Shettles had long hypothesized that this was not the case and for years had been suggesting that carbon dioxide gas could be used to safely immobilize living sperm, enabling researchers to select individual sperm cells for injection into the egg. Indeed, after the Belgian announcement, Dr. Shettles reiterated his earlier findings on this matter in a communication to the journal *Fertility and Sterility*. Dr. Van Steirteghem responded, in the same journal, that the Shettles approach could, in fact, prove useful in some applications of the new technique. All of this may move closer the day when procedures such as this are used for sex preselection, despite whatever controversy may attend that use.

Looking further ahead into the future, Dr. E. J. Leiberman, formerly of the National Institutes of Health, once suggested that women might eventually have "a special diaphragm that will let through only the sperm that carries, let's say, the male sex and hold back those that carry the female sex." Dr. Shettles did, in fact, report some years ago, in the *Journal of Urology*, that he had been able to devise filters that would permit passage of the male-producing sperm while restraining most of the female sperm. The differences in size, however, are very slight, and, further complicating matters, the differences vary from man to man. Filtration is still a possibility, but Dr. Shettles believes that chemical filtration may be more promising than mechanical filtration. Unfortunately, there seems to have been no further work on this in recent years.

Dr. Charles Birch, while head of the Sydney (Australia) University School of Biological Sciences, predicted that researchers would eventually come up with chemicals packaged in the form of pills that would determine sex. If male offspring were desired, Dr. Birch said, the husband would simply take one of the "little boy pills" just before intercourse, or, if a daughter was wanted, one of the "little girl pills." When you consider what such commonplace chemicals as vinegar, baking soda, and caffeine can do to the sperm, this idea does not seem so farfetched.

In fact, in 1980, various researchers noted that women who received the fertility drug Clomid give birth to an unusually high number of daughters. Dr. Shettles observed some years ago that women who take Clomid usually exhibit more acidic cervical secretions. The drug helps women ovulate, but it does not make their secretions hospitable to male-producing sperm. Clomid, in fact, Dr. Shettles reports, seems to produce just the kind of conditions that favor the female-producing sperm.

In an article in *Drug Therapy* (May 1976), Dr. Shettles called attention to this aspect of Clomid; the problem, he reported then, could be overcome by giving women the drug potassium iodide. This substance promotes production of cervical secretions such as exist at or near the time of ovulation. In his own practice, Dr. Shettles often uses potassium iodide in women who want to conceive boys but who have, for one reason or another, unusually acidic secretions. (Some forms of potassium iodide have been nonprescription in the past and may still be; however, you should not use this substance except on the advice of your own doctor.) Guaifenesin (see the "Questions and Answers" chapter) can also be useful in overcoming acidic conditions in women using Clomid.

Some have proposed using Clomid to help women conceive daughters, an idea that Dr. Shettles believes is medically irresponsible. Clomid is a powerful drug that affects the hormonal chemistry of body and brain; it often overstimulates the ovaries, so that multiple eggs develop and multiple births ensue. Multiple births are riskier than single births, both for the mother and for the babies. Moreover, many couples may not be financially prepared for multiple births. And since Clomid by no means guarantees daughters, you could end up with two, three, or even more sons—all at once. The Clomid data is important, nonetheless, because, in Dr. Shettles' opinion, it adds further support to his contention that acidic conditions within the female help "select" female offspring.

In 1995, researchers at Johns Hopkins, Duke University, and elsewhere reported that sperm have olfactory receptors that enable them to "sniff" or "smell" the egg—and that they may be drawn toward it by "chemoattraction." This finding excited a lot of attention in the media. Some scientists suggested that the discovery may lead to a new form of contraception, the idea being that if you can short-circuit the sperm cell's "sense of smell," it won't be able to find its way to the egg.

It is vaguely possible that male-producing and female-producing sperm have variant sensitivities to the egg's "perfume" and that these variations could also be exploited to impart a selective advantage to one or the other. A fanciful notion, perhaps, but not out of the question.

In the more distant future a number of exotic tools may contribute to sex selection. Already lasers are being tested, in animal experiments, to help visualize the two different types of sperm. When treated with special chemicals, sperm nuclei can be made to "light up" when they pass through a laser beam. Because one type contains more genetic material than the other, the amount of light differs, depending on whether it is male-producing or female-producing. Light meters can measure these flashes of light, as the sperm swim by, and computers can count the two types and distinguish between them. Such systems might eventually be coupled with other innovations that will shuttle those sperm that emit girl-fluorescence in one direction and those that emit boy-glow in the other.

Genetic engineering may also be applied to sex selection one day. Egg and sperm cells may be directly manipulated to ensure that they yield offspring of the sex desired. Genetic screening may be used to provide information on an individual's chances of conceiving children of one or the other sex. The same thing might be applied to couples to see what, in combination, they are most likely to produce in the way of boys and/or girls. Researchers have already found a particular genetic marker that can be isolated in the blood that may be predictive of a much higher than average probability of producing sons. (Don't run to your family doctor to have this test, however; it is not widely available at present and is used for other reasons.)

Researchers affiliated with the Massachusetts Institute of Technology reported that they believe they have found the "trigger"—a single gene—within the Y chromosome that determines male sex. It is the absence of this gene in the X chromosome that results in development of female offspring. As our knowledge of sexual differentiation moves from the chromosomal to the genetic level, we may eventually find entirely new ways of preselecting sex. Again, though, the availability of this technology at the "consumer" level is years, perhaps decades, away.

Someday it also may be possible to add—or subtract—genetic components, so that individuals will produce children of only one sex—at least until they request a genetic "adjustment." Such manipulations might be used to produce children not only of a specific sex but also those with specific eye, hair, and skin coloring, and so on. On the other hand, society may decide it wisest to disallow such developments. In any event, such capabilities will not exist for many, many years.

Whatever comes to pass, we hope that parents will always reproduce not because, first and foremost, they want a child of a particular sex but because they want and love children, irrespective of sex.

The 100 Most Popular
Boy and Girl Names

(And Some Alternatives)

Bertha and Henry among the most popular names? Well, they were in the 1880s. Times change and so do fashions in names. In the United States, the most popular girl name of all time is Mary. Mary held the number-one spot for an astonishing eight decades—and probably much longer. But beginning in the 1880s, when the government began reliably tabulating the popularity of names, Mary reigned supreme all the way up until the 1960s, when she was dethroned by the upstart Lisa, who was herself soon supplanted by the more durable Jessica. John was number one on the male side of the roster for decades, ultimately giving way to Robert, then James, and then Michael. Recently, Jacob and Emily have claimed the top spots. Incidentally, they were fairly popular back in the 1880s, as well!

Government and other statisticians, some working for big businesses, are continually revising lists of the most popular names in the country, using computer programs to help them crunch the numbers they obtain from birth registries all around the country. Some of them try to discern trends—what the most popular names are region by region, state by state, city by city—and others try, without much success, to figure out the psychology (and purchasing patterns) of people with different names. Sociologists want to know if particular names influence personality in any way, or success or popularity.

There's no denying that names can be fascinating. It is clear that the media have a big influence on name popularity these days. The names of popular soap opera characters, for example, have been found to crop up with increasing frequency in birth registries and then fade—in sync with the sinking fortunes of the same soap opera personas. It is also interesting to note that male names tend to change in popularity more slowly than do female names. Could this be because, as a society, we are a bit more rigid or traditional in our view of males? Or that men continue to insist on passing *their* names on to their male offspring?

Well, as you can see, it's easy to get lost in names—and prospective parents in particular, we have discovered, are fascinated by the name game. So here we offer the latest tabulations from our government registries, updated in early 2006.

The Most Popular Names by Rank

Rank	Male Name	Female Name
1	Jacob	Emily
2	Michael	Emma
3	Joshua	Madison
4	Matthew	Olivia
5	Ethan	Hannah
6	Andrew	Abigail
7	Daniel	Isabella
8	William	Ashley
9	Joseph	Samantha
10	Christopher	Elizabeth
11	Anthony	Alexis
12	Ryan	Sarah
13	Nicholas	Grace
14	David	Alyssa
15	Alexander	Sophia
16	Tyler	Lauren
17	James	Brianna

18	John	Kayla
19	Dylan	Natalie
20	Nathan	Anna
21	Jonathan	Jessica
22	Brandon	Taylor
23	Samuel	Chloë
24	Christian	Hailey
25	Benjamin	Ava
26	Zachary	Jasmine
27	Logan	Sydney
28	José	Victoria
29	Noah	Ella
30	Justin	Mia
31	Elijah	Morgan
32	Gabriel	Julia
33	Caleb	Kaitlyn
34	Kevin	Rachel
35	Austin	Katherine
36	Robert	Megan
37	Thomas	Alexandra
38	Connor	Jennifer
39	Evan	Destiny
40	Aidan	Allison
41	Jack	Savannah
42	Luke	Haley
43	Jordan	Mackenzie
44	Angel	Brooke
45	Isaiah	Maria
46	Isaac	Nicole
47	Jason	Makayla
48	Jackson	Trinity
49	Hunter	Kylie
50	Cameron	Kaylee
51	Gavin	Paige
52	Mason	Lily

53	Aaron	Faith
54	Juan	Zoë
55	Kyle	Stephanie
56	Charles	Jenna
57	Luis	Andrea
58	Adam	Riley
59	Brian	Katelyn
60	Aiden	Angelina
61	Eric	Kimberly
62	Jayden	Madeline
63	Alex	Mary
64	Bryan	Leah
65	Sean	Lillian
66	Owen	Michelle
67	Lucas	Amanda
68	Nathaniel	Sara
69	Ian	Sofia
70	Jesus	Jordan
71	Carlos	Alexa
72	Adrian	Rebecca
73	Diego	Gabrielle
74	Julian	Caroline
75	Cole	Vanessa
76	Ashton	Gabriella
77	Steven	Avery
78	Jeremiah	Marissa
79	Timothy	Ariana
80	Chase	Audrey
81	Devin	Jada
82	Seth	Autumn
83	Jaden	Evelyn
84	Colin	Jocelyn
85	Cody	Maya
86	Landon	Arianna
87	Carter	Isabel

88	Hayden	Amber
89	Xavier	Melanie
90	Wyatt	Diana
91	Dominic	Danielle
92	Richard	Sierra
93	Antonio	Leslie
94	Jesse	Aaliyah
95	Blake	Erin
96	Sebastian	Amelia
97	Miguel	Molly
98	Jake	Claire
99	Alejandro	Bailey
100	Patrick	Melissa

What is remarkable is how consistently people in states across the country pick the same names. Emma is number one in Alaska at the same time that she is the most popular in Alabama, Minnesota, New Hampshire, and elsewhere. There are a few exceptions. Sophia, for example, is the current favorite in the District of Columbia. Alyssa is number one in New Mexico, and Madison is at the top in Wyoming. There is currently a little more variation among the top-ranked male names, but Jacob is at the top in most states. William is number one in the District of Columbia, Joshua in Hawaii, Daniel in California, Ethan in Idaho, Michael in New York, and José in Texas.

If You Are in the Mood for Something Different

If these popular names are well, too popular, and you want something more unusual, consider these seldom-used names. For girls:

Adana
Blayke
Celestine
Dylan
Eris
Ferrin

Gwendaly
Honore
Ilsa
Jae
Kaylan
La-Tasha
Madigan
Oakley
Pepper
Quiarra
Racquell
Shadon
Tora
Ulinda
Victory
Wray
Xian
Ylana
Zandra

And for the boys:

A-Jay
Brik
Chevy
Dyson
Eden
Free
Goran
Holland
Ivy
Jovon
Klaus
L'mar
Macarthur

Nikko
Oron
Parish
Quinlan
Rafe
Skyelar
Treven
Urban
Vidas
Westin
Xackery
Yale
Zac

Write to Us!

We welcome your letters. Even though we can answer only a small proportion of the large volume of letters we receive, we appreciate receiving your input and your stories of success (and, occasionally, failure). We are always looking for interesting, instructive, and/or inspirational letters to include in future editions. Sometimes we run the letters themselves, other times we use what is related to us in the form of useful anecdotes in different parts of the book.

You will find a questionnaire on the next pages of the book. We urge you to fill it out and mail it back to us. Do not include your charts, but if you want to add more details about your experience using our method, please do so in the form of a letter attached to your questionnaire. We will not use your name without your permission.

Write to:
David M. Rorvik
P.O. Box 9281
Portland, OR 97207
USA

Reader Questionnaire

WE SUCCEEDED IN OUR GOAL TO CONCEIVE A
_____ (boy/girl).
WE DID NOT SUCCEED IN OUR GOAL TO CONCEIVE A
_____ (boy/girl).
Date we are filling out this questionnaire _____

Our (boy/girl) _____ was born on _____.
This was our (first, second, or ?) _____ child.
Gender of our other children _____
Wife's age _____ Occupation _____
Husband's age _____ Occupation _____

Have you ever tried sex-preselection methods before? _____
If yes, what method(s) did you use? _____
_____
When and with what results did you use the method(s) you describe
above? _____
_____
If you have used methods other than the Shettles method, how do you
compare those other methods with the Shettles method? _____
_____
Do you know other couples who have used the Shettles method?
_____
If so, with what results did they use the Shettles method? _____
_____
Do you know couples who have used methods other than the Shettles
method? _____ Which methods did they use? _____
_____

What were the results? _____

Have you heard of sex-preselection methods that are not discussed in this book? _____ If so, what are those methods? _____
_____

Was your pregnancy or delivery complicated in any way? _____ If yes, please specify. _____
_____

If you did not succeed in giving birth to a child of the sex you wanted, do you believe you carefully followed all of the recommendations in this book? _____ Why did you not succeed, insofar as you can tell?
_____

Did you use birth control in the twelve months prior to becoming pregnant in this attempt? _____ If so, what type of birth control and for how long a period? _____
_____

For how many menstrual cycles did you keep your basal body temperature (BBT) charts prior to making your effort? _____

For how many menstrual cycles did you keep cervical mucus (CM) charts prior to making your effort? _____

If you did not use the BBT method, explain why. _____
_____

If you did not use the CM method, explain why. _____
_____

Please check any of the following that you used instead of or in addition to the BBT and CM methods to help you determine your time of ovulation. Also, if you used any of these, please rate them poor/fair/good/excellent in terms of how useful they were in your case:

Tes-Tape _____ Rating: _____

Ovulation prediction kits _____ Rating: _____

Mittelschmerz _____ Rating: _____

Other (please specify) _____ Rating: _____

How confident were you that you had found the time of ovulation when you made the attempt that resulted in your pregnancy?

Very confident _____

Somewhat confident _____

Not too confident _____

Not confident at all but decided to try anyway _____

What factors contributed to your confidence or lack of it?

_____

_____

For how many menstrual cycles did you practice, irrespective of the ovulation method you used, prior to making the attempt that resulted in your pregnancy? _____

Does either side of the family have a preponderance of offspring of one gender or the other? _____ If yes, please explain._____

_____

Does the wife suffer from ulcers, acid stomach, or scanty cervical secretions? _____ If yes, please explain. _____

_____

Does the husband's job expose him to unusual stress, such as heat, radiation, high altitude, underwater pressures, noxious gases or other chemicals? _____ If yes, please specify. _____

_____

Does either husband or wife spend a lot of time at a computer terminal? _____ If yes, explain how much time. _____

_____

Have you ever seen any evidence that those who spend a lot of time at computer terminals or working around computers tend to have more female offspring? _____ If yes, please explain. _____

_____

(We have begun to hear reports, far from substantiated at this point, that computer workers may have more female offspring. We would particularly welcome input on this topic from people in the computer field or those who work with computers or at computer terminals for prolonged periods of time. Perhaps some of you could do informal surveys among your coworkers. Of course, the more specific you can be, the better.)

Has the husband ever had a sperm count performed? _____

If yes, what were the results (in numbers)? _____

When performed? _____

Does either husband or wife have any infertility problems (now or in the past)? _____ Explain. _____

_____

Have you ever tried any of the *in vitro* methodologies either to try to overcome an infertility problem or for sex preselection? _____ If yes, when, where, for what reason, and with what result? _____

_____

_____

Have you used any method to detect the sex of the baby you were pregnant with (either in this pregnancy or in an earlier one)? _____ If yes, when, why, and what, if anything, did you do with the information you obtained? _____

_____

Would you be interested in using these detection methods to prompt an abortion in the event the sex of the child you were carrying was not the gender you desired?

Very interested _____

Somewhat interested _____

Not very interested _____

Not interested at all _____

If medical scientists were able to alter the characteristics of the child you were carrying (to change eye or hair color, for example, gender, IQ, etc.), is this something you would want to avail yourself of?

Definitely _____

Possibly _____

Probably not _____

Absolutely not _____

If you answered "Definitely" or "Possibly," please enlarge on the circumstances under which you would find it desirable to make changes in the unborn while still in the womb. _____

_____

_____

If everybody could easily preselect the gender of their children, do you think this would be a good thing or a bad thing (and, briefly, why do you think this)? Yes _____ No _____

Why? _____

_____

_____

Did you find any aspect of the Shettles method distasteful, annoying, or distracting from sexual pleasure? _____ If yes, please explain.

_____

Did both husband and wife agree on the importance of using sex pres-election?

Yes, agreed equally _____

Wife wanted to use it more than husband _____

Husband wanted to use it more than wife _____

If one spouse had objections to the method, what were they?_____

_____

Was your doctor aware that you were using these methods?_____

If yes, was he/she approving? _____disapproving? _____

skeptical? _____ helpful? _____ indifferent? _____

Did you tell family and friends about your effort? _____

If yes, what was the reaction? _____

Did you have any religious qualms about using the method?_____

If yes, how did you resolve those qualms?_____

_____

Have you ever spoken to your clergyman about these methods? _____ If yes, what was his/her reaction? _____

_____

If you did not succeed in conceiving a child of the sex you desired, how would you rate your reaction?

Deep sorrow and marked disappointment _____

Moderate disappointment but no real sorrow _____

A little disappointment _____

No disappointment at all _____

If you did not succeed, do you plan to try again? _____

If yes, will you use the Shettles method again? _____

If you had not succeeded in conceiving a child of the sex you desired this time, would you have kept on trying until you did? In other words, how many more times would you have tried, if any, before giving up? _____

If you do *not* now regard your family as complete, how many more children do you want and of which sex? _____

How did you hear about this book? _____

_____

Is there anything in the book that you find confusing or believe needs clarification? _____

_____

_____

_____

Two sets of questions follow—one for those who wanted boys, one for those who wanted girls.

Questions to be answered only by those who were trying for a BOY:
On which day of the menstrual cycle in which you became pregnant did you *first* have intercourse? _____

On which days thereafter did you have intercourse? (Give each day of the cycle, up to the day of suspected ovulation.) _____ On which days of the cycle did you use condoms? _____ If you did not use condoms, explain what, if anything, you used for birth control. _____

_____

Did the intercourse that you believe resulted in pregnancy take place on the day of ovulation? _____ If you were using the BBT, did this intercourse take place the morning you believed you had recorded your last low temperature before the abrupt rise? _____ Later in the day on which you believe you noted this last low temperature? _____ Did it take place the next morning? _____ If it was the next morning, had the upward shift in temperature taken place by then? _____ At what time of day or night did this intercourse occur? _____

If you were using the CM, did you note the "peak" symptoms at the time of this intercourse? _____ If not, explain what your CM symptoms were at the time of the intercourse that you believe resulted in pregnancy. _____

_____

For about how long after this intercourse did your "peak" CM symptoms continue? _____ In general, for how long, during most of your cycles, do your "peak" symptoms persist (in days or hours)? _____ Do your CM symptoms tend to be "regular" and repetitive from one cycle to another? (If no, explain.) _____
_____

Did you use an ovulation prediction test kit to help you pinpoint your time of ovulation? _____ How long did you wait after you first detected the LH surge before you had intercourse in the attempt for the boy? _____ Were you testing twice a day with the kit? _____ For how long, if at all, after the day of suspected ovulation did you abstain from intercourse? _____ Did you use condoms when you resumed intercourse? _____
Did the husband avoid tight-fitting clothing prior to the attempt?

_____

Did he drink coffee prior to the attempt? _____
Did the wife have orgasm? _____ If yes, did this precede the husband's orgasm?_____ Coincide with it? _____ Follow it?

_____

Was penetration from the rear? _____
Was deep penetration at the time of male orgasm achieved?

_____

Is there, on either side of the family, a predominance of one sex among offspring? _____ If so, explain. _____
_____

Did you both want a boy with equal intensity, as far as you can tell?
_____ If no, which of you wanted a boy more? _____
Why did you want a boy? _____
_____
_____

Questions to be answered only by those who were trying for a GIRL:
How many times did you have intercourse from the first day of your cycle up to the cutoff date prior to ovulation? _____
Did you use contraception on any of those days? _____

If yes, why did you use contraception? _____

_____

At what time of day did you last have intercourse on the cutoff date?

_____

Did you schedule the cutoff date a certain number of days from the probable time at which the *last low* temperature of the cycle would be recorded or from the day on which the *upward* temperature shift would probably be noted? _____

Did you initially schedule the cutoff date *three days* before suspected ovulation? _____

Did you succeed in becoming pregnant on this schedule? _____

If yes, how many attempts (cycles) did it take before you became pregnant? (Do not count practice cycles in which no serious effort was made to become pregnant.) _____

If you did not succeed in becoming pregnant with a three-day cutoff, how many times did you try before giving up and going to a shorter cutoff schedule? _____

If you didn't succeed at three days before ovulation, at what interval did you make the next attempt: two and one half days? two days? _____

Did you succeed using *this* cutoff interval? _____ If yes, how many cycles did it take before you became pregnant? _____

_____

If no, at what interval *did* you finally become pregnant (whether with a boy or girl)? _____

How long did you abstain from intercourse *after* the cutoff date?

_____

When you resumed sexual relations, did you use contraception? _____ If yes, what type did you use? _____

If you used the CM method (either alone or in conjunction with BBT), what were your CM symptoms on the last day on which you had intercourse prior to ovulation? _____

_____

In general, do your CM symptoms tend to be regular and repetitive from cycle to cycle? (If no, explain.) _____

_____

How soon after your last intercourse—prior to ovulation—did your CM symptoms "peak"? _____

Did you use one of the ovulation prediction test kits in your effort? _____

If so, which one? _____

For how many cycles and for how many days did you use the kit? _____

Did you test with the kit once or twice a day? _____

Did you find the test kit useful in helping you predict when you would ovulate and, if so, precisely how did you find it useful in your effort to conceive a girl? Please be specific and give us details, as we need more information on these kits when applied to sex selection. _____ _____ _____ _____

Did the wife experience orgasm during any episode of intercourse during any attempt to achieve a girl? _____

Was the face-to-face position used during intercourse? _____

Did the husband succeed in shallow penetration at the time of orgasm? _____

Is there, on either side of the family, a predominance of one sex among offspring? _____ If so, explain. _____ _____

Did you both want a girl with equal intensity, as far as you can tell? _____ If no, which of you wanted a girl more? _____

Why did you want a girl? _____ _____

We may use information you have provided us in this questionnaire in the next edition of our book but will not use real names or addresses. Thank you for your help. Please use the space below for general comments.

_____ _____

Bibliography

Unsigned Reports

Boy or girl, take your pick. *Science Digest,* April 1974.

Caffeine vs. fertility. *Reader's Digest,* October 1994.

Children of divers found to be predominantly female. *OBGYN News,* October 15–31, 1982.

Choosing your baby's gender. cbsnews.com, November 7, 2002.

Coital patterns successful in predicting child's sex. *OBGYN News,* February 15, 1971.

Disease and changes in the sex ratio. *Research in Reproduction,* April 2, 1982.

Fetal sex preselection: technology/prospects/issues. *OBGYN Topics,* September-October 1987.

Girls from space. *Discover,* March 1988.

Male sperms swifter off the mark. *New Scientist,* July 20, 1967.

New reproductive technologies. *ACOG Technical Bulletin,* March 1990.

Of coitus and the baby's sex. *Medical World News,* August 13, 1972.

Physician guides couples in "selecting" their children. *Las Vegas Review-Journal,* October 15, 1990.

Popular baby names. Social Security Online, www.ssa.gov, January 7, 2006.

Predetermining the sex of your child. BabyHopes.com, 2005.

"Preselecting" infant's sex may help prevent birth defects. *OBGYN News,* October 15, 1975.

Psychotomimetics affect sex ratio? *OBGYN News,* December 15, 1970.

Regulation of LH rhythms. *Research in Reproduction,* July 1990.

Separation of X and Y spermatozoa. *Research in Population,* January 1974.

Sex determination study viewed skeptically. *OBGYN News,* April 1, 1979.

Sex selection for social reasons unlikely to skew gender balance. Press release, European Society for Human Reproduction and Embryology, September 24, 2003.

Sex selection popular among infertile women. Press release, University of Illinois at Chicago, March 10, 2005.

Sperm shape whips up a storm. *Medical World News,* August 12, 1960.

Success reported in predicting and influencing fetal sex. *OBGYN News,* March 15, 1977.

Toward sex on order. *Time,* June 27, 1960.

Two shapes of sperm found in humans. *New York Times,* June 5, 1960.

Would sex selection skew balance? Reuters, October 31, 2003.

X-Y sperm separation. *Andrology Newsletter,* Spring 1995.

Signed Reports

Bailey, R. Will sex selection create a violent world without women? reasonline.com, October 6, 2004.

Beernink, Ferdinand J., et al. Sex preselection through albumin separation of sperm. *Fertility and Sterility* 59 (1993): 382–86.

Benendo, Franciszek. The problems of sex determination in the light of personal observations. *Polish Endocrinology* 21 (1970): 200–7.

Bennett, D., and Boyse, E. A. Sex ratio in progeny of mice inseminated with sperm treated with H-Y antiserum. *Nature* 246 (November 30, 1973): 308–9.

Bhattacharya, B. C. Sex control in mammals. *Zeitschrift für Tierzuchtung und Zuchtungsbiologie* 72 (1958): 250–54.

———. The different sedimentation speeds of X and Y sperm and the question of optional sex determination. *Zeitschrift für Wissenschaftliche Zoologie* 166 (1962): 203–50.

Bouton, K. New light on male infertility. *This World,* July 11, 1982.

Brandriff, Brigitte F., et al. Sex chromosome ratios determined by karyotypic analysis in albumin-isolated human sperm. *Fertility and Sterility* 46 (1986): 678–85.

Carroll, R. T. Too good to be true—Dr. Eugen Jonas Method. skepdic.com, July 18, 2005.

Check, Jerome H., et al. Improvement of cervical factor with guaifenesin. *Fertility and Sterility* 37 (1982): 707–8.

Check, William. ACOG meeting combines scientific and social perspectives. *OB/GYN World,* November 1987.

Corson, Stephen L. Self-prediction of ovulation using a urinary luteinizing hormone test. *Journal of Reproductive Medicine* 31 (August 1986): 760–63.

Dahl, Roald. Ah, sweet mystery of life. *New York Times,* September 14, 1974.

Damewood, Marion D. Medical procedures and psychosexual evaluation for in vitro fertilization. *Clinical Practice in Sexuality* 3 (1989): 14–21.

Dawson, E. R. *The Causation of Sex in Man.* 2nd ed. London: H. K. Lewis & Co., 1917.

Dickens, B. M. Preimplantation genetic diagnosis and "savior siblings." *International Journal of Gynaecology and Obstetrics* 88 (1) (January 2005): 91–96.

Edwards, R. G., and Gardner, R. L. Sexing of live rabbit blastocysts. *Nature* 214 (May 6, 1967): 576–77.

Ericsson, R. J.; Langevin, C. N.; and Nishino, M. Isolation of fractions rich in human Y sperm. *Nature* 246 (December 14, 1973): 421–24.

Etzioni, A. Sex control, science and society. *Science* 161 (September 13, 1968): 1107–12.

———. Selecting the sex of one's children. Letter to the editor, *Lancet* 1 (May 11, 1974): 932–33.

Everhardt, Ellen, et al. The effects and fertility rationale of vaginal sodium bicarbonate douching. *Infertility* 13 (1990): 35–51.

France, J. T., et al. A prospective study of the preselection of the sex of offspring. *Fertility and Sterility* 41 (1984): 894–900.

Freedman, D. S.; Freedman, R.; and Whelpton, P. K. Size of family and preferences for children of each sex. *American Journal of Sociology* 66 (1960): 141–46.

Frenkiel, Nora. Would you choose the gender of your next child? *The Oregonian* (New York Times News Service), November 16, 1993.

Galton, Laurence. Parent and child: choosing the sex of a child. *New York Times Magazine,* June 30, 1974.

Garcia, Jairo E. Gamete intrafallopian transfer (GIFT): historic perspective. *Journal of in Vitro Fertilization and Embryo Transfer* 8 (1991): 1–4.

Gilbert, V. Low-tech ways to choose your baby's gender. preconception.com, 2004.

Gordon, M. J. Control of sex ratio in rabbits by electrophoresis of spermatozoa. *Proceedings of the National Academy of Sciences* 43 (1957): 913–18.

Gottlieb, S. U.S. doctors say sex selection acceptable for non-medical reasons. *British Medical Journal* 323 (2004) (no. 7317): 828.

Guerrero, R. Association of the type and time of insemination within the menstrual cycle with the human sex ratio at birth. *New England Journal of Medicine* 291 (November 14, 1974): 1056–59.

Hammond, Mary G. Monitoring ovulation. *Contemporary OB/GYN* (Special Issue), September 1987.

Han, Tie Lan, et al. Detection of X- and Y-bearing human spermatozoa after motile sperm isolation by swim-up. *Fertility and Sterility* 60 (1993): 1046–51.

Haney, Daniel Q. What makes baby a boy or girl? Scientists may have the answer. *The Oregonian* (Associated Press), December 23, 1987.

Harlap, S. Gender of infants conceived on different days of the menstrual cycle. *New England Journal of Medicine* 300 (June 28, 1979): 1445–48.

Harsanyi, Z., and Hutton, R. *Genetic Prophecy: Beyond the Double Helix.* New York: Rawson-Wade Publishers, 1981.

Hart, D., and Moody, J. D. Sex ratio: experimental studies demonstrating controlled variations—preliminary report. *Annals of Surgery* 129 (May 1949): 550–71.

Hatzold, Otfried. Personal communication, Munich, Germany. September 3, 1974.

Iritani, Akira. Current status of biotechnological studies in mammalian reproduction. *Fertility and Sterility* 50 (1988): 543–51.

James, W. H., Cycle day of insemination, coital rate and sex ratio. *Lancet* 1 (January 16, 1971): 112–14.

———. Sex ratios in large sibships, in the presence of twins and in Jewish sibships. *Journal of Biosocial Science* 7 (April 1975): 165–69.

Janerich, D. T. Sex ratio and season of birth. Letter to the editor, *Lancet,* April 24, 1971.

Kaiser, R.; Broer, K. H.; Citoler, P.; and Leister, B. Penetration of spermatozoa with Y-chromosomes in cervical mucus by an in vitro test. *Geburtshilfe und Frauenheilkunde* 34 (June 1974): 426–30.

Keynes, R. D. The predetermination of sex. *Advancement of Science* 24 (1967): 43–46.

Kleegman, S. J. Therapeutic donor insemination. *Fertility and Sterility* 5 (1954): 7–31.

————. Can sex be planned by the physician? In *Fertility and Sterility,* Proceedings of the 5th World Congress on Fertility and Sterility (Stockholm, June 16–22, 1966), B. Westin and N. Wiquist, eds. Amsterdam: Excerpta Medica (International Congress Series No. 133, 1967): 1185–95.

Knaack, J. Arbitrary influence on sex by sedimented bull sperms—results of a large-scale test. *Fortpflanzung Besamung und Augzucht der Haustiere* 4 (1968): 279–82.

Kristof, N. D. China: officials warn of shortage of future wives. *The Oregonian* (New York Times News Service), July 21, 1993.

Krzanowski, M. Dependence of primary and secondary sex-ratio on the rapidity of sedimentation of bull semen. *Journal of Reproduction and Fertility* 23 (1970): 11–20.

Laino, C. Vitamins may enhance sperm. *Medical Tribune,* November 23, 1995.

Langendoen, S., and Proctor, W. *The Preconception Gender Diet.* New York: M. Evans and Company, 1982.

Lappé, M. Choosing the sex of our children. *The Hastings Center Report* 4 (February 1974): 1.

Leff, D. N. Boy or girl: now choice not chance. *Medical World News,* December 1, 1975.

Levine, R. J., et al. Differences in the quality of semen in outdoor workers during summer and winter. *New England Journal of Medicine* 323 (1990): 12–16.

Levy, Jacob. The surplus of male births among Jews—a contribution to the question of sex-determination. *Koroth* 6 (November 1973).

Lobel, Susan M., et al. The sex ratio of normal and manipulated human sperm quantitated by the polymerase chain reaction. *Fertility and Sterility* 59 (1993): 387–92.

Lowe, C. R., and McKeown, T. The sex ratio of human births related to maternal age. *British Journal of Social Medicine* 4 (1950): 78–85.

Martin, L. Your child's sex—can you choose? *Parents,* October 1981.

McCartney, D. Why sex selection should be legal. *Journal of Medical Ethics* 27 (2001): 302–307.

Ng, Soon-Chye, et al. Micromanipulation: its relevance to human in vitro fertilization. *Fertility and Sterility* 53 (1990): 203–19.

Novitski, E., and Kimball, A. W. Birth order, parental ages and sex of offspring. *American Journal of Human Genetics* 10 (1958): 268–75.

Parkes, A. S. Mythology of the human sex ratio. In *Sex Ratio at Birth—Prospects for Control,* C. A. Kiddy and H. D. Hafs, eds. Champaign, Ill.: American Society of Animal Science, 1971: 38–42.

Pearson, P. L.; Geraedts, J. P. M.; and Pawlowitski, I. H. Chromosomal studies on human male gametes, in *Proceedings of the Symposium on Chromosomal Errors in Relation to Reproductive Failure,* A. Bove and C. Thibault, eds. Paris: Centre International de l'Enfance (September 1973): 219–27.

Perez, A., et al. Sex ratio associated with natural family planning. *Fertility and Sterility* 43 (1985): 152–53.

Perlman, D. Choosing sex of livestock. *San Francisco Chronicle,* November 24, 1982.

Repetto, R. Son preference and fertility behavior in developing countries. *Studies in Family Planning* 3 (April 1972): 70–76.

Rhine, S. A.; Cain, J. L.; Cleary, R. E.; et al. Prenatal sex detection with endocervical smears: successful results utilizing Y-body fluorescence. *American Journal of Obstetrics and Gynecology* 122 (May 1975): 155–60.

Roberts, A. M. *Nature* 238 (1972): 223.

Robinson, D.; Rock, J.; and Menkin, M. F. Control of human spermatogenesis by induced changes of intrascrotal temperature. *Journal of the American Medical Association* 204 (April 22, 1968): 80–87.

Rohde, W.; Porstmann, T.; and Dörner, G. Migration of Y-bearing spermatozoa in cervical mucus. *Journal of Reproduction and Fertility* 33 (April 1, 1973): 157–69.

Rohde, W.; Porstmann, T.; Prehn, S.; and Dörner, G. Gravitational pattern of the Y-bearing human sperm in density gradient centrifugation. *Journal of Reproduction and Fertility* 42 (March 1975): 587–91.

Ruben, David. We'll have one of each, please. *Parenting,* October 1990.

Saltus, R. Why parents turn to sex selection. *San Francisco Sunday Examiner and Chronicle,* January 30, 1983.

Schellen, A. *Artificial Insemination in the Human.* Amsterdam: Elsevier Publishing Co., 1957.

Schilling, E. Sedimentation as an approach to the problem of separating X and Y chromosome bearing spermatozoa. In *Sex Ratio at Birth: Prospects for Control,* C. A. Kiddy and H. D. Hafs, eds. Champaign, Ill.: American Society of Animal Science, 1971: 76–84.

Schröder, V. Sex control of mammalian offspring and the biochemical and physiological properties of X and Y bearing sperm. *Animal Breeding Abstracts* 10 (December 1942): 252.

Schwegel, Janet. *The Baby Name Countdown.* New York: Marlowe & Company, 1996.

Seguy, B. Methods of natural and voluntary selection of the sexes. *Journal de Gynécologie Obstétrique et Biologie de la Reproduction* 4 (1975): 145–49.

Shettles, L. B. Nuclear morphology of cells in human amniotic fluid in relation to sex of infant. *American Journal of Obstetrics and Gynecology* 71 (1956): 834–38.

––––––. Biological sex differences with special reference to disease, resistance and longevity. *Journal of Obstetrics and Gynecology of the British Empire* 65 (1958): 288.

––––––. Cervical factors in reproduction. *American Journal of Obstetrics and Gynecology* 14 (November 1959): 635–43.

––––––. Observations on human spermatozoa. *Bulletin of the Sloane Hospital for Women* 6 (1960): 48.

––––––. Nuclear morphology of human spermatozoa. *Nature* 187 (1960): 254.

––––––. Differences in human spermatozoa. *Fertility and Sterility* 12 (1961): 20.

––––––. Head differences in human spermatozoa. *Journal of Urology* 85 (1961): 355.

––––––. Human sperm populations. *International Journal of Fertility* 7 (1962): 175.

––––––. The great preponderance of human males conceived. *American Journal of Obstetrics and Gynecology* 89 (May 1, 1964): 130–33.

———. Factors influencing sex ratios. *International Journal of Gynaecology and Obstetrics* 8 (September 1970): 643–47.

———. Use of the Y chromosome in prenatal sex determination. *Nature* 230 (March 5, 1971): 52–53.

———. Sperm morphology, cervical milieu, time of insemination and sex ratios. *Andrologie* 5 (1973): 227–30.

———. Sex selection. Letter to the editor, *American Journal of Obstetrics and Gynecology,* February 15, 1976.

———. Human spermatozoa filtration. *Journal of Urology* 116 (1976): 462–63.

———. Potassium iodide enhancement of cervical and uterine secretions. *Drug Therapy* (May 1976).

———. Sex determination of offspring. *Medical Aspects of Human Sexuality* (March 1977).

———. Differentiation of X and Y chromosomes in sperm (reply to Dr. Glass). Letter to the editor, *American Journal of Obstetrics and Gynecology,* May 15, 1977.

———. Cells in cervical mucus reveal sex of infant. *American Journal of Obstetrics and Gynecology* (January 15, 1978).

———. Why more "accidental" births are girls. *Medical Aspects of Human Sexuality* (February 1978).

———. Letter to the editor, *Fertility and Sterility* 29 (March 1978): 368.

———. X and Y spermatozoa. *Infertility* 2 (1979): 81–87.

———. Birth-sex ratio in vitro conceptions. Letter to the editor, *American Journal of Obstetrics and Gynecology,* May 15, 1982.

———. Is sex selection frivolous? No. *Physicians Weekly,* January 19, 1987.

———. Letter to the editor, *OB/GYN World,* May 1987.

———. How sperm commence movement and their isolation for in vitro fertilization and sex selection. *American Journal of Obstetrics and Gynecology* (July 1990).

———. Tubal embryo successfully transferred in utero. *American Journal of Obstetrics and Gynecology* (December 1990).

———. Carbon dioxide and sperm motility. *Fertility and Sterility* 61 (1994): 576.

———. Fraternal twin conceptions. *Contemporary OB/GYN* (May 1994).

———. Sperm sniff out eggs. *Hopkins Medical News* (Fall 1995).

Simcock, B. W. Sons and daughters—a sex preselection study. *Medical Journal of Australia* (1985): 541.

Simpson, J. L., and Carson, S. A., The reproductive option of sex selection. *Human Reproduction* 14 (1999): 870–872.

Singh, Mukul; Brij, B. Saxena; and Premila, Rathnam. Clinical validation of enzymeimmunoassay of human luteinizing hormone in the detection of the preovulatory luteinizing hormone surge in urine. *Fertility and Sterility* 41 (February 1984): 210–17.

Smits, L. J. M., et al. Time to pregnancy and sex of offspring: a cohort study. *British Medical Journal* 331 (2005): 1437–1438.

Stolkowski, J., and Lorrain, J. Preconceptional selection of fetal sex. *International Journal of Gynaecology and Obstetrics* 18 (1980): 440–43.

Sugiyama, S. Personal communication, December 28, 1982.

Sullivan, Walter. New way devised to pick child's sex. *New York Times,* September 23, 1987.

Talwar, P. P. Effect of desired sex composition in families on the birth rate. *Journal of Biosocial Science* 7 (April 1975): 133–39.

Teitelbaum, M. S. Factors associated with the sex ratio in human populations. In *The Structure of Human Populations,* G. A. Harrison and A. J. Boyce, eds. London: Oxford University Press, 1972: 90–109.

Thorne, R. W. Choosing child's sex a possibility in the '80s with new techniques. *OBGYN News,* February 1–14, 1982.

Turner, Barbara Kay. *Baby Names for the '90s.* New York: Berkley Books, 1991.

Unterberger, F. Sex determination and hydrogen ion concentration. *Deutsche Medizinische Wochenschrift* 58 (May 6, 1932): 729–31.

Van Steirteghem, A. C., et al. Letter to editor. *Fertility and Sterility* 61 (1993): 576.

Vear, C. S. Preselective sex determination. *Medical Journal of Australia* 2 (1977): 700–2.

Verny, T., and Kelly, J. *The Secret Life of the Unborn Child.* New York: Summit Books, 1981.

Wachtel, S. S.; Koo, G. C.; Zuckerman, E. E.; Hammerling, U.; Scheid, M. P.; and Boyse, E. A. Serological crossreactivity between H-Y (male) antigens of mouse and man. *Proceedings of the National Academy of Sciences* 71 (April 1974): 1215–18.

Wakim, P. E. Determining the sex of baby rabbits by ascertaining the pH of the vagina of the mother before mating. *Journal of the American Osteopathic Association* 72 (October 1972): 173–74.

Walensky, Loren D., et al. Odorant receptors and desensitization proteins colocalize in mammalian sperm. *Molecular Medicine* 1 (1995): 130–41.

Walters, LeRoy. Is sex selection frivolous? Yes. *Physician's Weekly,* January 19, 1987.

Wang, Huai-Xiu, et al. Assessment of the separation of X- and Y-bearing sperm on albumin gradients using double-label fluorescence in situ hybridization. *Fertility and Sterility* 61 (1994): 720–26.

Weil, R. Fear and fertility in Las Vegas. *Omni,* December 1982.

Westoff, C. F.; Potter, R. B., Jr.; and Sagi, P. C. *The Third Child: A Study in the Prediction of Fertility.* Princeton, N.J.: Princeton University Press, 1963.

Westoff, C. F., and Rindfuss, R. R. Sex preselection in the United States: some implications. *Science* 184 (May 10, 1974): 633–36.

Whelan, E. *Boy or Girl?* Indianapolis: Bobbs-Merrill Company, 1977.

Wilcox, A. J., et al. Timing of sexual intercourse in relation to ovulation. *New England Journal of Medicine* 333 (1995): 1517–21.

Williamson, N. E., et al. Evaluation of an unsuccessful sex-selection clinic in Singapore. *Journal of Biosocial Science* 10 (1978): 375–88.

Wilson, M. A. *The Ovulation Method of Birth Regulation.* New York: Van Nostrand Reinhold Company, 1980.

Wolinsky, Howard. Drugstores will sell kit to help select baby's sex. *Chicago Sun-Times,* August 3, 1986.

Young, J. Martin. *How to Have a Boy; How to Have a Girl* (2 volumes). Houston: Young Ideas Publishing, 1995.

Zarutskie, P. W., et al. The clinical relevance of sex selection techniques. *Fertility and Sterility* 52 (1989): 891–905.

Zirkle, C. The knowledge of heredity before 1900. In *Genetics and the 20th Century: Essays on the Progress of Genetics During Its First 50 Years,* ed. L. Gunn. New York: Macmillan, 1951: 35–37.

Index

About the Authors

LANDRUM B. SHETTLES, M.D., Ph.D., was an associate professor of obstetrics and gynecology at Columbia University's College of Physicians and Surgeons and director of research at the New York Fertility Foundation. He is the author of *From Conception to Birth: The Drama of Life's Beginnings*.

DAVID M. RORVIK is a former *Time* magazine science and medicine reporter and has contributed articles to the *New York Times Magazine, Harper's, Reader's Digest*, and other magazines.